PLANNING FOR SOLAR ENERGY:
Promoting Solar Energy Use Through Local Policy and Action

TABLE OF CONTENTS

CHAPTER 1

Solar Energy Use as a
Local Planning Issue

David Morley, AICP, and Brian Ross

 For centuries humans have designed buildings and settlements to take advantage of light and heat from the sun. While many of these design techniques fell out of favor with the advent of fossil-fuel-produced heat and electricity, in recent years communities across the U.S. and throughout the world have taken a renewed interest in both passive and active solar energy use. In many industrialized nations, rising fuel prices and concerns over energy security during the 1970s planted the first seeds of the modern market for solar energy production. However, these initial investments in solar technology remained quite modest until the first decade of the twenty-first century.

Solar-friendly federal and state policies and financial incentives have driven much of this recent solar market growth. The reasons for this support are straightforward. Sunlight is a safe, clean, and abundant energy source available throughout the U.S. Solar energy systems on the rooftops of homes and businesses lower energy bills and provide jobs for system installers and manufacturers. Solar power plants generate electricity without the emissions and pollution associated with fossil fuels.

When local plans and regulations fail to explicitly address solar energy use, it can create a significant barrier to adoption and implementation of solar technologies. The purpose of this report is to provide planners, public officials, and other community stakeholders with a comprehensive guide to planning for solar energy use.

Planners are dedicated to helping communities chart courses to more sustainable futures, finding the right balance of new development and essential services, environmental protection, and innovative change. Ideally, they promote policies and practices that improve equity, strengthen economies, and enhance natural systems. In order to be effective, planners must think comprehensively and act strategically.

While there are numerous resources discussing strategies for growing local solar markets, this report is distinct in its emphasis on the planning perspective. This perspective places the goal of supporting solar energy use within the context of a series of key community decision points about future growth and change.

This introductory chapter begins with a simple concept: Solar energy is a community resource and should, therefore, be treated as such. Next, this chapter introduces the five strategic points of intervention that planners, public officials, and other community stakeholders can use to foster opportunities for solar energy use and evaluate solar development opportunities. The chapter concludes with an overview of the report's goals, structure, and content.

SOLAR ENERGY AS A LOCAL RESOURCE

The concept of conventional energy reserves (such as our nation's oil, gas, or coal reserves) is readily understood by most planners. Similarly, at the local level, planners routinely assess their communities' economic, natural, and social (or human) resources in order to set priorities and make planning and

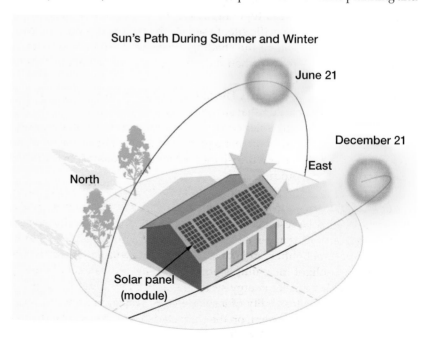

Figure 1.1. In the Northern Hemisphere, a south-facing roof maximizes solar access.

U.S. Department of Energy

development decisions. But what about our communities' solar resources, the solar energy "reserves" available for development? While every community in the U.S. has solar resources, very few consider how planning and development decisions affect the future availability of local solar resources or opportunities for private-sector solar development.

The Solar Resource

Solar irradiance, or solar radiation, refers to the electromagnetic energy that emanates from the sun. Because solar radiation can be harnessed to produce heat and electricity, it makes sense to look at it through the lens of resource management. While every location on earth receives some amount of solar radiation, a number of temporal, atmospheric, and geographic variables affect the quantity and quality of the solar resource available at a particular location.

The effects of time of day and season. The rotation of the earth means that the local solar resource varies throughout the day, with peak potential at midday when the sun is at its highest point in the sky. However, the local solar resource also varies by season. Because the earth's axis is tilted approximately 23.4 degrees to its orbital plane, the northern hemisphere is closer to the sun during the period between the spring and fall equinoxes.

The effects of atmospheric conditions. Air molecules, clouds, water, and particulate matter (including pollution and dust) all limit the amount of solar radiation that reaches the surface of the earth at any given time and location. Sunlight that passes through these obstructions has less energy than unobstructed solar irradiance. A number of factors influence local atmospheric conditions, including altitude, weather patterns, and the prevalence of human activities that produce pollution (e.g., driving and heavy industry).

The effects of latitude and the local landscape. Because the earth is spherical, solar radiation hits the surface of the earth at different angles throughout the year based on latitude. As the angle of entry increases, the amount of atmosphere the sunlight must pass through increases. In other words, lower latitudes receive more solar radiation throughout the year than higher latitudes. At the site level, though, the local landscape has a much greater effect on the solar resource available to a particular location. Structures, vegetation, and topography can limit the amount of solar radiation that reaches a specific site. Based on the factors discussed above, in the northern hemisphere the shading effects of these local landscape features are most profound when they are located on the adjacent south side of the site (Figure 1.1).

Measuring the solar resource. The term *insolation* refers to the amount of radiant energy from the sun that strikes a given surface area over a period of time. When discussing solar energy use, the two most common insolation metrics are kilowatt-hours per square meter (kWh/m²), which relates to electricity production, and British thermal units per square foot (Btu/ft²), which relates to heat production.

How much insolation is enough? Despite the local variation in the quantity of insolation, every community in the U.S. has opportunities to take advantage of the solar resource (Figure 1.2, p. 4). All but the most thoroughly shaded sites can, with proper site and building design, use solar radiation to enhance natural lighting and space heating. Sites with unobstructed access to direct sunlight for multiple hours each day, regardless of latitude, are often suitable for solar energy systems that produce heat or electricity. At the site level, the feasibility of a solar energy system in a specific location depends, to a large extent, on the characteristics of the local landscape referenced above (Figure 1.3, p. 5).

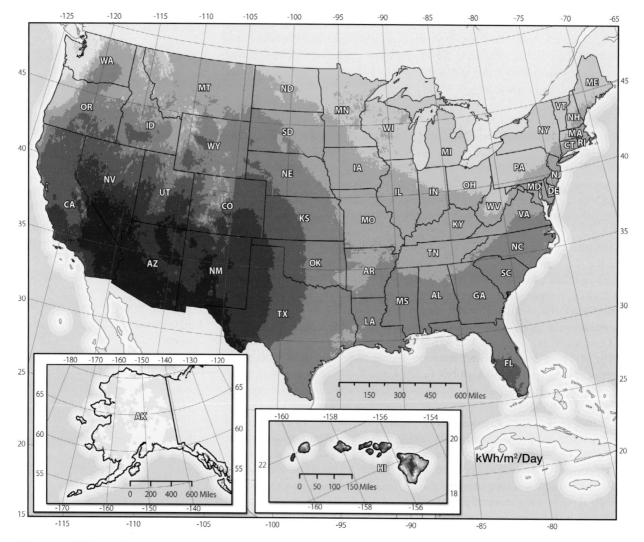

Figure 1.2. *This map shows national solar photovoltaics (PV) resource potential for the United States.*

Source: National Renewable Energy Laboratory 2013a

kWh/m²/Day

> 65	5.0 to 5.5	3.5 to 4.0
6.0 to 6.5	4.5 to 5.0	3.0 to 3.5
5.5 to 6.0	4.0 to 4.5	< 3.0

Land-Use Implications

Plans and regulations written without the solar resource in mind can limit solar energy use. Basic zoning concepts such as setbacks and height and lot coverage restrictions affect solar resource use. If the solar resource on a residential lot is limited to the front yard, a prohibition on accessory structures in the front yard will preclude solar development. Doing so is not necessarily the wrong decision (every community has its own aesthetic), but the decision should be deliberate, not inadvertent.

As with all developable resources, the harvesting of solar resources has land-use implications. Solar development has tradeoffs with other types of development and with the functioning or harvesting of other types of community resources.

For rooftop solar energy systems, land-use issues primarily relate to visual impacts, particularly when the solar resource requires installations that extend above the peak of pitched roofs or installations in areas with design standards or historic resources. For freestanding systems subordinate to a principal structure or use, considerations include visual impacts, stormwater management, and bulk and massing issues. For smaller primary-use freestanding systems, land-use issues can include compatibility with adjacent

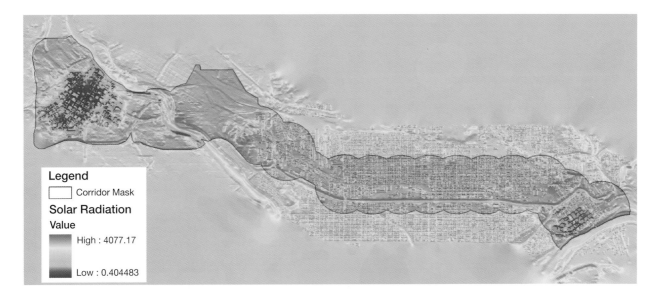

uses, visual impacts, stormwater management, and protection of natural systems. For large solar farms, the land-use implications may be analogous to those related to large-scale clean industrial uses, including compatibility with nearby uses, stormwater management, natural systems protection, access control, and adequate public facilities.

Figure 1.3. This map shows how local landscape features affect solar insolation along the light-rail corridor connecting the downtowns of Minneapolis and St. Paul, Minnesota.

CR Planning

Environmental Implications

Local solar energy use provides a number of environmental benefits for local communities. Solar energy is a carbon-free, emission-free, local fuel, which can help communities meet goals for greenhouse gas reduction, energy independence, and state or local renewable portfolio standards (see Chapter 5).

As with many planning-related issues, the environmental implications of solar energy use transcend jurisdictional borders, and the direct environmental benefits may accrue outside the community. These broader benefits include decreased emissions from centralized fossil-fuel power plants, slower expansion of fossil-fuel mining or drilling operations, and reduced water consumption for cooling towers at power plants.

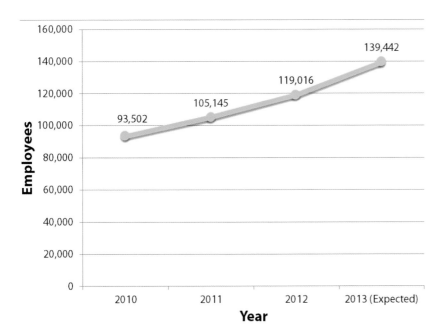

Figure 1.4. In 2012 there were approximately 119,000 solar workers in the U.S., a 13.2 percent increase over employment totals in 2011.

Source: Solar Foundation et al. 2012

Economic Implications

Solar energy use can help meet economic development goals as well. Approximately half of the jobs within the U.S. solar industry are local installation-related jobs (Solar Foundation and BW Research Partnership et al. 2012) (Figure 1.4, p. 5). Moreover, most solar installation work is retrofitting existing buildings, and is therefore not tied to the market for new construction. The expansion of local solar installation jobs thus offers opportunities for existing contractors to diversify into new markets.

Solar energy use also provides local economic value through import substitution. Import substitution is the use of a local resource in place of a nonlocal resource. Money spent on local resources stays within the local economy; money spent on nonlocal resources leaves the local economy. Since few communities have a coal mine or natural gas field, there is often a local economic benefit to substituting local solar energy for a power produced at a centralized plant using nonlocal fossil fuels.

FIVE STRATEGIC POINTS OF INTERVENTION

Communities that are ready to begin planning for solar energy use often wonder where to start. In fact, sometimes the idea of planning itself can seem diffuse to public officials, business leaders, community groups, and residents. One way to think about how planners and the local planning system can evaluate opportunities for solar energy use is through the five strategic points of intervention (Figure 1.5). These are points where planning process participants translate ideas into intentions and intentions into actions.

Visioning and Goal Setting

Community visioning is often the first step in developing any type of community plan, establishing new regulations and incentives, participating in development work, and making public investment decisions. Whether part of a planning process or on its own, visioning allows communities to identify new opportunities and priorities—including those related to solar energy. Often the explicit goal of a visioning exercise is to produce long-term goals and objectives. Planners then use the ideas and preferences expressed in visioning exercises to develop policies and action items for the community.

Once a community agrees upon its long-range goals and objectives, public officials should look to these goals and objectives when reviewing development proposals, making budget decisions, and performing other related tasks. These visioning meetings also provide the first and best opportunity for residents and other stakeholders to discuss how solar energy use connects to other community goals and values.

Plan Making

When stakeholders identify solar energy as a priority during visioning and goal-setting exercises, they are influencing the types of plans a community undertakes as well as what will be incorporated into existing plans in the future. Communities may choose to address solar energy use through a variety of plan types, including comprehensive plans, subarea plans, and functional plans, such as climate action plans, energy plans, or sustainability plans.

Local plans provide opportunities for communities to document existing conditions of local solar markets as well as how energy use relates to other community goals and priorities. Plans also typically establish goals and policies and lay out action steps for meeting those goals. These goals, policies, and action steps guide decision makers when making future decisions and may address topics like solar access protection, incentives, or preferential locations for new solar development.

Regulations and Incentives

Planners help communities develop and amend regulations and incentives that have an important influence on what, where, and how things get built and the preservation of land and buildings. Additionally, review boards with discretionary power rely on this information when making decisions about specific development proposals.

Promoting solar energy use through regulations involves identifying and removing unintended barriers and enacting appropriate standards for solar development in zoning, subdivision, and building codes. Communities can also use development regulations, administrative processes, and financial tools to incentivize solar energy use.

Development Work

Planners and other stakeholders in the planning process often have opportunities to influence the outcomes of development or redevelopment projects. One of the most important ways that planners can influence solar energy use is by providing information and assistance to interested property owners and developers considering the installation of solar energy systems. Additionally, local governments can use development agreements and discretionary approval processes, such as site plan or conditional use reviews, to advocate for the inclusion of solar in the development program for a site. Finally, localities are frequently involved in a variety of public-private partnerships and redevelopment projects, including mixed use developments, brownfields development, downtown revitalization, affordable housing, and transit-oriented developments—all of which have the potential to integrate solar energy systems.

Public Investments

Local governments undertake major investments in infrastructure and community facilities that support private development and improve quality of life. Public buildings and structures—such as city halls, libraries, schools, parking garages, and police and fire stations—all have solar potential. Installation of solar systems at these locations can help communities meet their energy-reduction goals as well as substantially reduce energy costs and energy cost uncertainty for the community over the long term. Furthermore, communities have opportunities to make direct investments in economic development and educational programs that support solar energy use.

OVERVIEW OF THE REPORT

Cities and counties have tremendous opportunities to address barriers to solar energy use through their plans and plan implementation tools. This report has three primary goals: to provide planners, public officials, and other community stakeholders with a basic rationale for planning for solar energy use; to summarize the fundamental characteristics of the U.S. solar market as they relate to local solar energy use; and, to explain how planners, public officials, and other community stakeholders can take advantage of the five strategic points of intervention to promote solar energy use.

The policies and approaches discussed herein are all rooted in contemporary practice. This is good news for any community with a new or renewed interest in promoting expanded use of its local solar resource because there are many examples of communities that have already taken bold steps in this direction.

After this introduction, Chapter 2 provides some background on the U.S. solar market, including a basic explanation of the range of technologies that fall under the umbrella of solar energy use, a brief summary of the role that pricing trends and utility policies play in solar market growth, and a general overview of the different scales and contexts for solar development.

The next five chapters take a closer look at each of the strategic points of intervention outlined above. Chapter 3 discusses why solar energy belongs on the local policy agenda and offers some tips to help planners guide community conversations about setting and balancing goals and priorities related to solar energy use. Chapter 4 offers specific guidance to help communities integrate solar-supportive goals, objectives, policies, and actions into local comprehensive, subarea, and functional plans. Chapter 5 explains the importance of calibrating development regulations and incentive programs to implement solar-related goals and policies. Chapter 6 summarizes some key strategies for promoting solar energy use through development services and public-private partnerships. Chapter 7 summarizes how local governments can make direct investments in solar through installations on public facilities and through economic development and educational programs.

The report concludes with Chapter 8, which reiterates the key themes discussed in the preceding chapters and touches on some emerging trends that may affect community efforts to promote solar energy use in the future. Finally, the report appendices contain a checklist to help communities evaluate how well their existing policy frameworks support solar energy use, a tool to help communities work through the steps of drafting new regulations for solar development, and examples of solar-supportive plan policies and development regulations from communities across the country.

CHAPTER 2

Overview of the U.S. Solar Market

Chad Laurent, Jayson Uppal, David Morley, AICP, and Justin Barnes

 Before initiating any visioning or goal-setting process related to solar energy use, it is helpful to have an understanding of the different technologies as well as the economic and policy drivers that influence the U.S. solar market. This chapter will provide planners and community stakeholders with an overview of commercially available solar technologies, summarize solar market trends, explain different financing mechanisms, review government and utility policies that support solar energy use, and contrast solar development at different scales and in different contexts.

TECHNOLOGIES

Several solar technologies convert sunlight into a practical form of energy, most commonly for electricity or heat. These technologies vary significantly in their costs, benefits, and access requirements. Passive technologies use site design and nonmechanical building materials to capture or direct heat or light. The discussion below refers to passive solar design as a blanket term for a wide range of ancient and modern passive technologies. In contrast, active solar technologies use electrical or mechanical equipment to convert solar irradiance to heat or electricity or to transport this newly converted energy. The discussion below focuses primarily on the three most prevalent active solar technologies: solar photovoltaic systems, solar thermal systems, and concentrating solar power systems.

Passive Solar Design

Passive solar design is a low-cost and simple way to mitigate a building's energy needs. This technique uses both site and building design to maximize the lighting and space heating benefits of solar radiation available at a specific location. While passive solar design is beneficial on its own, some design elements can also be used to enhance the performance of solar energy systems.

At the site scale, the goal of passive solar design is to maximize the amount of direct sunlight available to each building. Because building orientation often depends on street and lot layout, solar site design aims to orient streets and front lot lines in new subdivisions along an east-west axis, to the maximum extent possible (Figure 2.1).

For individual buildings, window orientation, material selection, and ventilation control are key elements of passive solar design. In colder climates, windows facing in the southern direction can capture solar energy during the day for space heating and natural light within the structure. In warmer climates, roof overhangs for southern-facing windows allow for indirect sunlight while preventing the heating effect. Solar design also takes advantage of materials that absorb and retain heat, which if implemented correctly can both absorb solar heat and disperse it within the building in cold seasons and absorb heat from inside the house in warm seasons, resulting in a cooling effect. Ventilation control allows structures to adjust the internal temperature with outside air, particularly useful in warmer climates where there are significant temperature fluctuations between daytime and nighttime.

Figure 2.1. This site plan for a residential subdivision shows how streets, lots, and buildings can be configured to maximize solar access.

Source: Erley and Jaffe 1979

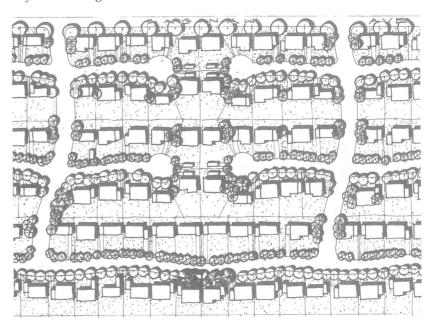

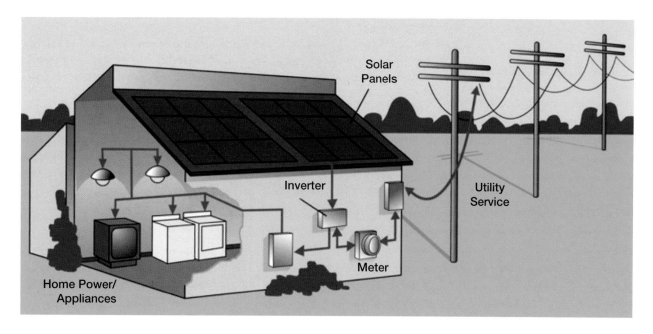

Solar Photovoltaic Systems

Solar photovoltaic (PV) systems use photovoltaic cell technology to harness radiant energy from the sun and create electricity. These cells are often packaged into panels that can be placed on rooftops or mounted on the ground. The cells can also be incorporated into other building materials such as roofing materials, facades, and even glass—an arrangement that is commonly known as "building-integrated PV." Since most PV cells produce direct current, or DC, electricity, an inverter is often added to the system to convert the output to alternating current, or AC, electricity, the form of electricity that most appliances and the electricity grid are designed to use.

Generally speaking, PV systems must receive direct solar radiation for multiple hours each day, on average, in order to generate enough electricity to justify installation costs. With that said, the specific thresholds for adequate electricity production depend on the placement and specifications of the system and how the electricity will be used. The National Renewable Energy Laboratory's (NREL) PVWatts calculator allows users to see how system parameters—such as size, array type, and tilt and azimuth angle—affect system performance (NREL 2013a).

One of the foremost benefits of PV technology is its flexibility to accommodate a range of uses. PV has been successfully implemented on homes, businesses, manufacturing facilities, and even utility-scale projects comparable in size with large fossil-fuel generation facilities. It is this flexibility that has contributed to its success in the market; over 7.7 gigawatts (GW) of PV capacity has been installed in the U.S., enough to power 1.2 million homes (SEIA and GTM Research 2013).

Solar Thermal Systems

Solar thermal systems harness the sun's thermal energy to heat a fluid, such as water or antifreeze, to satisfy hot water or space heating and cooling needs for residential, commercial, or industrial facilities. The technology is relatively simple: water or another heat-transfer fluid passes through panels or tubes that, when placed in sunlight, capture and transfer the radiant heat. These systems are typically mounted on the roof of the facility, and they may feed a hot water tank, heat exchanger, or thermally driven chiller (for solar cooling).

As with PV systems, solar thermal systems require access to direct sunlight. While locations that receive an average of at least 4.5 kWh/m²/

This diagram shows components of a typical residential grid-connected PV system.

U.S. Department of Energy

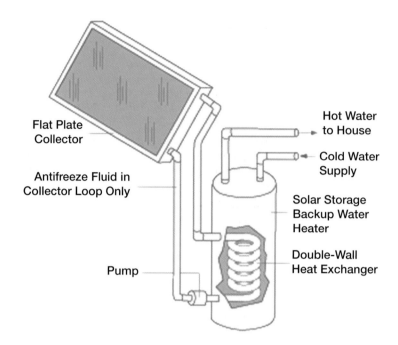

The fundamental components of an active, closed-loop solar thermal water-heating system

U.S. Department of Energy

day of insolation are ideal, sites that receive less solar radiation may still be suitable if local utility rates are high (Walker 2012). NREL's Federal Energy Management Program maps show how insolation, energy rates, and incentives affect the economics of solar thermal systems (NREL 2013b).

Thanks to the simplicity of the technology, solar thermal systems are one of the lowest-cost and most effective ways to capture solar energy. Each year, approximately 30,000 property owners in the U.S. choose to install solar thermal systems (SEIA 2013).

Concentrating Solar Power Systems

Concentrating solar power (CSP) systems use mirrors to focus light and heat a contained substance such as molten salts or water to create steam. These mirrors may be arranged as a trough focusing the light on a substance travelling through a tube, or as a dish focusing the light on a single point. The heat from that substance is harnessed to drive a mechanical engine, which subsequently drives an electric generator.

Parabolic trough concentrating solar power (CSP) collectors capture the sun's energy with large mirrors that reflect and focus the sunlight onto a linear receiver tube.

U.S. Department of Energy / National Renewable Energy Laboratory

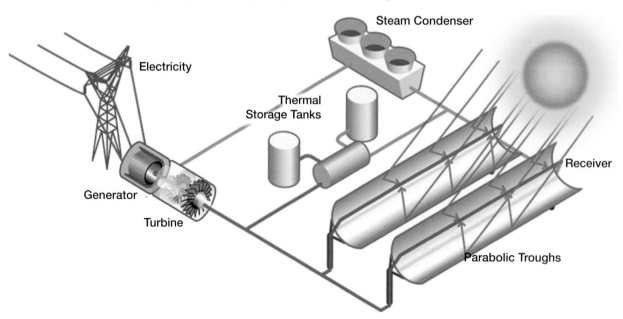

Unlike PV systems, CSP systems are generally only commercially viable on a large scale, typically for large industrial facilities or as a wholesale electricity generator for utilities larger than 100 megawatts (MW) in capacity. In order to meet this large scale, CSP systems require a significant amount of land, normally five to 10 acres per MW. Furthermore, CSP systems, like all thermal power plants, use large amounts of water (SEIA 2010).

Perhaps the primary benefit of CSP systems over PV systems in utility applications is that CSP systems can store energy more efficiently. While PV electricity production drops off substantially in the late afternoon and early evening, when electricity use is still high, the thermal energy collected by a CSP system can be stored for extended periods of time, allowing it to generate electricity as the sun sets. So far, CSP systems represent just a small fraction of the total installed solar capacity in the U.S., but over 800 MW of CSP projects are expected to come online in 2013 (Trabish 2012).

COSTS AND FINANCING

The upfront costs associated with solar energy use vary considerably based on technology, scale, and the availability of financial incentives. If a developer plats a subdivision in accordance with passive solar design principles, there may be little to no added premium for improved natural lighting and space heating. However, some passive solar building-design features and all solar energy systems do involve upfront costs in excess of similar developments without those features.

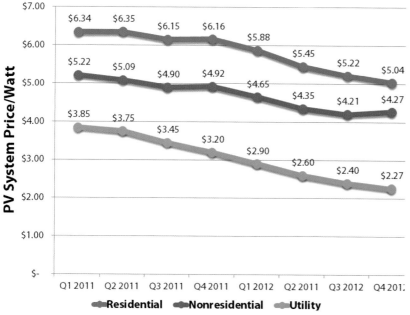

Figure 2.2. *Average installed PV system price/watt by market segment, 2011–2012.*

Source: SEIA and GTM Research 2013

Solar Energy System Trends

As solar PV technology has matured over the past decade, residential installation costs have dropped from $12 per watt in 1998 to about $5 per watt today (Figure 2.2). Between 1998 and 2011, installation costs in the U.S. market have declined on average five to seven percent per year (Barbose et al. 2012). The declining cost trend has accelerated in the past few years, falling 27 percent in 2012 alone. This decline can be attributed primarily to tumbling prices for PV equipment, particularly PV modules, which fell 47 percent in 2012 to $0.68 per watt (SEIA and GTM Research 2013).

NEW TECHNOLOGIES

There are a number of technological developments that may have a profound effect on solar energy use in the coming years. These new technologies may affect not only the equipment costs and space requirements associated with solar energy production, but also help to address the challenge of integrating an intermittent source of electricity with the power grid.

Unlike silicon-based solar cells, dye-sensitized solar cells offer transparency, low cost, and high power-conversion efficiencies under cloudy and artificial light conditions.

Wikimedia Commons / Sastra

Improvements in crystalline-silicon, thin-film, multijunction III-V, organic, and dye-sensitized photovoltaics will, sooner or later, likely make it practical to integrate PV systems into a wide range of building materials, hardware, and even apparel. If and when this happens, traditional rooftop solar energy systems may play a much less significant role in the U.S. solar market.

Electricity generated from solar irradiance fluctuates based on environmental conditions, and these fluctuations pose a problem for the existing electrical grid, which was designed to move a steady flow of electricity. The key to overcoming this barrier may be found in emerging storage technologies, such as high-energy batteries, compressed air energy storage, and thermal energy storage (Lichtner et al. 2010).

It is important to note that this trend is more dramatic for larger-scale installations than for small residential systems. This is due, in part, to the fact that "soft" costs (e.g., local permitting fees) comprise a larger percentage of total costs for small systems than for large installations and that these costs have not decreased at the same pace as equipment prices.

As a result of these declining costs, solar PV is increasingly becoming a financially viable electricity source. In some European countries with strong markets, solar power is already nearing the point at which the cost of electricity from a PV installation is equal to or less than the retail rate. This point is commonly referred to as "grid parity," and the U.S. Department of Energy expects that this milestone can be reached across the U.S. by as early as 2020 (Ong et al. 2012). Grid parity for PV systems is measured using total system costs over the life of the system, and most of these costs are associated with the initial capital investment by the PV system owner.

Solar thermal is a mature technology, and equipment costs are closely tied to the prices of raw materials such as copper, glass, and aluminum. The industry experienced rapid growth between 1978 and 1985, coinciding with the availability of a 40 percent federal investment tax credit offered in response to the 1970s oil crisis. Since then, growth in solar thermal installations has been modest, and the U.S. currently ranks 36th in the world in per capita installed capacity (SEIA 2013).

As of 2011 the installed costs of CSP systems globally ranged from about $4.60 per watt to $10.50 per watt, with most of the variation attributable to the amount of thermal energy storage capacity (IREA 2012). By early 2012, the installed capacity of CSP systems in the U.S. exceeded 520 MW, but high upfront costs and the scarcity of willing financiers have limited CSP market growth (SEIA 2012).

In terms of installed capacity, recent gains from utility-scale solar farms have far outstripped gains from residential installations. Between 2009 and 2012, utility-scale farms in the U.S. added 2,866 MW of capacity, while residential systems added 1,200 MW (SEIA and GTM Research 2013). This growth in large installations is largely attributable to state renewable portfolio standards (RPS).

An RPS requires that a certain amount of the energy that utilities and energy suppliers use to serve their customers come from renewable energy resources within a certain time frame (e.g., 25 percent by 2025). In some cases, RPS policies create further specific requirements for solar energy or distributed generation, usually referred to as "carve-outs." Utilities and energy suppliers comply with an RPS by purchasing and retiring renewable energy certificates (RECs). A REC serves as proof that one megawatt-hour (MWh) of electricity has been generated from a renewable resource, and can often be bought and sold separately from the associated electricity. The term solar REC, or SREC, refers to electricity generated using a solar resource. Some states permit solar thermal systems to generate SRECs, translating thermal energy generated to an equivalent amount of electricity.

As with all energy technologies, financial incentives continue to play a vital role in driving the growth of the U.S. solar market. Some incentives provide capital for an upfront investment in a solar installation, commonly in the form of a rebate or a grant. Other incentives pay out over time, based on system performance. Finally, in addition to the federal investment tax credit on all PV installations valued at 30 percent of the upfront cost, many state and local governments offer credits that may be used to offset income, sales, or property tax liability. See Chapter 5 for more in-depth discussion of financial incentives.

Financing Mechanisms

For many interested in taking advantage of the benefits that solar energy can provide, the high upfront equipment and installation costs may be prohibitive to successfully developing a solar facility.

A number of mechanisms are available to residents, businesses, and governments to spread the investment cost over a period of time, reducing the upfront burden. The two most prevalent mechanisms are debt financing and third-party ownership.

Debt Financing. An owner may choose to borrow money to cover a portion or the full upfront cost of a solar installation and pay that back over time with interest. This may be a loan from a private institution or, in the case of projects for public entities, the debt may be in the form of a bond. In this model, the borrower retains complete ownership over the project, is eligible to receive financial incentives, and is responsible for the operations and maintenance of the facility.

Third-Party Ownership. Under a third-party ownership arrangement, a private solar developer will build, own, and operate a solar PV or thermal system on behalf of a host customer. The developer then either leases the system back to the host customer, sells the heat or electricity through a power purchase agreement (PPA), or, in the case of a thermal system, charges a fee based on avoided energy cost under the terms of a performance contract. With third-party ownership, hosts avoid upfront and operating costs, and solar developers may retain access to financial incentives.

Third-party ownership is quickly becoming the most popular way to finance PV projects. In 2012, between 60 and 90 percent of new PV projects were third-party financed in the top solar markets, including California, Massachusetts, Colorado, and Arizona (U.S. Department of Energy 2012). Because governments and nonprofits are tax exempt, these entities often find third-party ownership to be particularly beneficial, as the private developer can take advantage of the federal investment tax credit and pass the savings along through the PPA or lease.

Currently, third-party ownership for solar thermal systems is rare. Most third-party solar thermal projects use PPAs, where the host purchases heat instead of electricity, and are sited on nonresidential buildings (SEIA and GTM Research 2012).

In some states, third-party PV system ownership falls within the definition of a public utility, thus facing regulation like any other utility. Figure 2.3 shows which states explicitly allow third-party PPAs, which states explicitly disallow third-party PPAs, and in which states the policies are unclear. Furthermore, some tax incentives, rebates, and net metering arrangements (see discussion below) may be unavailable to third-party owners.

Figure 2.3. *At least 22 states and the District of Columbia and Puerto Rico authorize or allow third-party solar PV purchase power agreements.*

Source: DSIRE 2013a

Note: This map is intended to serve as an unofficial guide; it does not constitute as legal advice. Seek qualified legal expertise before making binding financial decisions relating to a 3rd-party PPA.

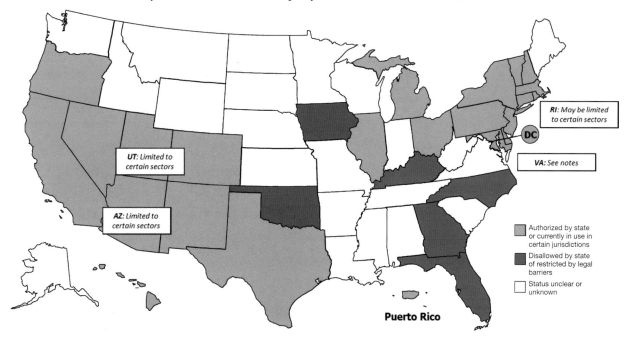

UTILITIES

State and local regulation of utilities affects local solar-market potential in significant ways. There are three major types of utilities: investor-owned utilities (IOUs), municipal utilities (munis), and cooperatives (coops). Each utility type has its own unique regulatory structure and business models that affect how solar is allowed to connect to the grid and how much the utility is willing to pay for the power produced.

Investor-Owned Utilities

Investor-owned utilities (IOUs) are electric utility companies that issue stock and are owned by their shareholders. IOUs can range in size from small local companies serving a few thousand customers to multinational corporations which serve millions of customers.

The IOU model is the dominant electricity utility structure not only in the U.S., but worldwide. IOUs are regulated monopolies that have exclusive rights to sell electricity with a guaranteed rate of return in a service territory in exchange for agreeing to provide nondiscriminatory access and safe, reliable, and reasonably priced electricity. An IOU also has an obligation to make a profit for its shareholders, and shareholders may live outside the IOU's service territory.

Traditionally, IOUs have been vertically-integrated monopolies that own generation, transmission, and distribution facilities within their service territories. IOUs generally have large customer bases and serve large geographic areas.

There are 194 IOUs in the U.S. serving roughly 98 million customers (APPA 2013) (Figure 2.4). All IOUs in the U.S. are regulated by either state or local regulatory agencies (often called a public service commission or public utility board). In addition, the Federal Energy Regulatory Commission regulates interstate wholesale electricity sales and interstate transmission which affects IOUs that operate across multiple states.

Regulatory agencies control the following aspects of an IOU's business operations:

• Rate base, or the assets the utility can earn a profit on

• Rate of return, or how much profit can be earned

• Resource planning, or how generation resources are acquired

• Rates and rate design

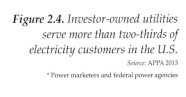

Figure 2.4. *Investor-owned utilities serve more than two-thirds of electricity customers in the U.S.*
Source: APPA 2013
* Power marketers and federal power agencies

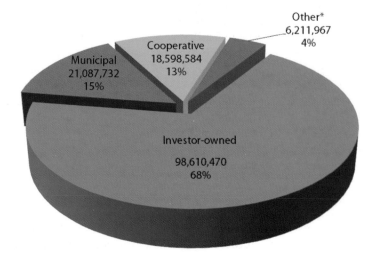

Other*
6,211,967
4%

Cooperative
18,598,584
13%

Municipal
21,087,732
15%

Investor-owned
98,610,470
68%

This regulated rate structure is based on "cost-of-service," in which the total costs to serve customers, including returns to investors, are administratively determined and then collected from ratepayers. Customer-sited solar energy systems and net metering (see discussion below) decrease the amount of electricity purchased by utility customers. Lower utility returns can result in a lower credit rating and a higher cost of capital for the utility, which can result in higher costs to ratepayers. As a result of this structure many utilities may be incentivized to avoid or prevent the development of customer-sited solar (Shirley and Taylor 2009).

Municipal Utilities

Municipal utilities (munis) are owned and operated by a municipality, city, or town. There are just over 2,000 munis in the U.S. serving 20.9 million customers, ranging in size from small municipal distribution companies that serve a few thousand customers to large systems that serve millions of customers, such as the Los Angeles Department of Water and Power and the Long Island Power Authority (APPA 2013). Munis are generally financed by municipal treasuries and revenue bonds.

The structures of municipal utilities vary across the U.S. and even within states. Some munis own generation facilities, while others purchase power from independent power producers or federal power marketers such as the Tennessee Valley Authority.

Munis are regulated by local governments, public utility boards, or in some cases by state public utility commissions. Munis can be run by local city councils, independent boards elected by voters or city officials, public utility districts, or special state authorities. This locally based regulatory structure has a number of implications for PV development.

Because local residents and business owners are the "shareholders," munis may have the flexibility to pursue broader community or public interest goals if directed to do so. However, over 50 percent of munis serve populations of 10,000 or fewer, meaning they have small staffs and can lack the resources necessary to implement advancements in billing, smart meters, renewable generation, and other technology or business structure innovations. Due to the relatively small number of ratepayers of many munis, the cost implications of technology and PV development can have more visible impacts on the electricity prices of ratepayers. Nevertheless, as most municipal utilities are exempt from some federal energy regulations, they have opportunities to implement innovative solar policies such as feed-in tariff programs (see Chapter 5).

Rural Electric Cooperatives

Rural electric cooperatives (coops) are nonprofit utilities owned by the members that are served by the coop. Coops provide electric service to rural or semi-rural areas that are often underserved by IOUs or munis. There are 874 cooperatives that serve 18.5 million customers in the U.S. (APPA 2013). Decisions are made through an elected board of directors. Some coops own the distribution network, power lines, and power plants, while others only own the distribution network and purchase wholesale power from third parties. Coops tend to have a small number of employees and serve a relatively small number of customers. Coop customers can be located across large geographic areas, making access to transmission and substation infrastructure challenging. As with munis, technology upgrades and investments can be relatively expensive due to the small number of ratepayers paying the cost of those upgrades.

Net Metering

Net metering policies allow electricity customers to take credit for the energy that their PV system produces and, as a result, pay an electric bill based only

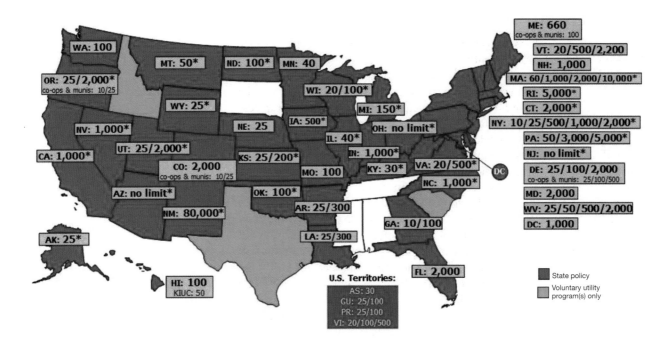

Figure 2.5. *Currently, 43 states and the District of Columbia have enabled net metering.*

Source: DSIRE 2013a

Note: Numbers indicate individual capacity limit in kW. Some limits vary by customer type, technology and/or application. Other limits might also apply. This map generally does not address statutory changes until administrative rules have been adopted to implement such changes.

on the net amount of electricity that the customer had to purchase from the grid during a billing period. Net metering is often visually depicted as the customer's electric meter running backwards, allowing the customer to offset purchases of electricity from the grid with exports to the grid at a different time. The exact structure of the net metering policy varies by state or even by utility, and state policies vary widely in terms of a number of key policy-design issues, including system size, eligible customer type (e.g., residential or commercial), treatment of excess generation at the end of a billing period, and ownership of renewable energy credits or solar renewable energy credits (DSIRE 2013a) (Figure 2.5).

Some programs allow customers to take credit for excess energy produced and apply that credit to later electric bills. For example, a PV system is likely to produce more energy in the summer, and a customer could apply the excess credit to electric bills in the winter when the PV system is producing less electricity. Electricity bills have many different components, including charges for transmission maintenance and distribution of electricity, plus the cost of the actual energy (i.e., the electrons themselves). Under most net metering policies, the customer receives the full retail value for at least some portion of the electricity produced by the PV system—thus providing the potential to offset a customer's entire electric bill. The retail value of electricity is often significantly higher than only the cost of the energy; therefore, net metering policies allow the customer to receive much more value for the energy produced by a PV system than if the system only received wholesale electricity rates or only offset the energy portion of the bill. This allows customers to receive a faster payback or better return on investment. This is especially true in parts of the country with high retail electricity rates.

While net metering is strongly supported by most solar industry advocates, as noted above many IOUs worry about how net metering affects their bottom lines. Since net metering decreases the amount of power purchased from the utility without necessarily reducing the costs associated with distribution infrastructure or maintenance, growth in net metering may, in some instances, lead to rate increases. Furthermore, utility regulators in many states are cautious about making policy changes without more information about how the expansion of net metering is likely to affect utilities and utility customers. As a consequence, regulators, utilities, and solar advocates are engaged in a lively debate over appropriate net-metering policy and ratemaking (SEPA 2013).

Interconnection

Interconnection refers to the technical and procedural requirements necessary to safely, reliably, and efficiently connect an electricity-generating system (e.g., a PV system) to the electricity grid (Varnado and Sheehan 2009). In order for a PV system to net meter, rather than rely on batteries to store the electricity, the system must be interconnected. The interconnection process sets forth guidelines and criteria in order to allow electricity to flow from the PV system out into the grid.

Traditionally, utilities (regardless of the type) owned generation facilities and thus had control over the how producers connected to their electricity grid systems (Varnado and Sheehan 2009). Interconnection procedures run counter to this established method by allowing a potentially large number of electricity-generating systems to interconnect at various points along a grid. Without established interconnection procedures, the cost of studying the potential impacts of connecting to the grid could overwhelm the cost of a PV system (Alderfer et al. 2000). Therefore, it is critical that utilities use well-established guidelines and best practices to facilitate the interconnection of PV systems in order to safely and efficiently allow and capture the benefits of PV generation. The Interstate Renewable Energy Council's Model Interconnection Procedures establishes a recommended process for interconnecting PV systems using established safety and reliability standards including those of the Institute of Electronics and Electrical Engineers and Underwriters Laboratories, Inc. (IREC 2009).

Energy Deregulation

Electricity and natural gas utilities have historically both sold and distributed energy to their customers, meaning they operate as monopolies in their respective service areas. However, since the 1990s, more than two dozen states have taken steps to separate the sale of energy from its distribution (USEIA 2009 and 2012). This is commonly referred to as energy deregulation. In deregulated or partially deregulated states, qualifying customers have the choice to purchase gas or electricity from energy supply companies instead of their local utilities. Regardless of where the customer chooses to purchase energy, the local utility still delivers the gas or electricity and maintains the distribution infrastructure.

Electric utility deregulation has a number of potential implications for the solar industry and for customers interested in purchasing solar energy. In states with deregulated electric utilities, solar power producers can market electricity directly to retail customers without worrying about violating a protected utility monopoly. And some deregulated states allow municipalities to create aggregation programs to combine the purchasing power of all retail customers. The municipality may then choose to purchase solar power on behalf of all aggregated customers. Finally, in deregulated markets net metering becomes, potentially, more complex. This is because there may be three, instead of two, affected parties: the customer, the utility, and the energy supplier (Barnes and Varnado 2010).

SCALES AND CONTEXTS

Solar technologies are highly scalable. Developers and property owners can employ passive solar design techniques for a single home or an entire master planned community. Solar PV and thermal systems are modular, consisting of any number of individual collectors, and space requirements vary from a few dozen square feet for residential installations to tens, hundreds, or even thousands of acres for the largest PV systems. While CSP systems are also scalable, as mentioned above, this technology is typically most cost-effective when deployed at a large scale (i.e., hundreds of acres).

Because of their scalability, solar energy systems can either be accessory to an existing primary use or structure or the primary land use for a particular

Figure 2.6. *This rooftop system in Longmont, Colorado, meets about 25 percent of the household's electric needs and provides back-up power to critical appliances during utility power outages.*

Altair Energy (NREL 08880)

Figure 2.7. This small PV system provides on-site power to a home in Westcliffe, Colorado.

Warren Gretz (NREL 10599)

Figure 2.8. *Florida Power & Light Company's DeSoto Next Generation Solar Energy Center, in rural DeSoto County, Florida, is a 25-MW solar power plant.*

SunPower Corporation (NREL 23816)

parcel. Accessory systems supply heat or power on-site, while primary-use systems provide electricity to the grid or nearby properties.

Solar Energy Systems as an Accessory Land Use

An accessory use is typically defined as a secondary activity incidental to the primary use of the property. In most cases an accessory solar thermal or PV installation is on a home or business where the primary use of the property is either household living or a commercial activity. In these cases, the system utilizes roof space or other space on the property to provide heat or electricity for the primary use of the property (Figure 2.6). Accessory system sizes can vary widely depending on the size of the structure or the amount of land available on a given property. Consequently, accessory systems may be as small as a few panels on a residential roof or as large as thousands of panels on the roof of a factory, warehouse, or big box store.

Apart from rooftop installations, freestanding accessory solar energy systems can be ground- or pole-mounted in any open area on a given property (Figure 2.7). For example, a freestanding solar thermal system may provide heat to a swimming pool, while a freestanding PV system may be accessory to a farm or, in the case of PV-topped parking lot shade structures, to a large retail development.

Solar Energy Systems as a Primary Land Use

A primary land use is the main purpose for which a site is developed and occupied. While large "solar farms" may be the most familiar form of a solar PV or CSP system as a primary land use, smaller primary-use PV installations (sometimes referred to as "solar gardens") are popping up in communities across the country. Regardless of size, primary-use systems are typically freestanding, and the principal economic function of the land hosting a primary-use system is producing solar power for off-site consumption.

Utility- or wholesale-size solar farms may have rated capacities of multiple megawatts and may cover tens to thousands of acres of land. These installations primarily supply power for offsite consumption through the electrical grid (Figure 2.8). Meanwhile, in many cases smaller installations provide electricity either directly or virtually (through a virtual net-metering arrangement with a local utility) to nearby public facilities, residences, or businesses.

Solar industry insiders and advocates often use the term "community solar" or "community-based solar" to refer to projects where community members own shares in the PV system, can subscribe to receive the generated solar power, or can purchase the output of an off-site PV system to offset their own electricity bills (Coughlin et al. 2012). Many community solar projects are smaller-scale (e.g., less than 2 MW) primary-use installations (Figure 2.9).

Figure 2.9. *The 345.6 kW Brewster Community Solar Garden, located on a former municipal sand pit and dump site in Brewster, Massachusetts, provides enough electricity to power dozens of homes.*

Brewster Community Solar Garden Cooperative, Inc.

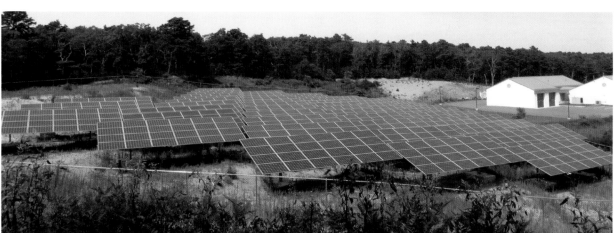

SUMMARY

There are a number of different types of solar technologies that convert sunlight into practical forms of energy, and these technologies vary significantly in their costs, benefits, and solar access requirements. Planners and other community stakeholders interested in promoting solar energy use would be wise to take note of how scale, context, utility policy, and the availability of incentives affect the viability of different solar technologies. These variables, ultimately, establish the limits of any local effort to support solar energy use through planning.

GLOSSARY OF KEY TERMS RELATED TO SOLAR DEVELOPMENT

The ideas and strategies discussed in this report are rooted in the following defined terms. It is important to note that other sources may use some of these terms differently, and the intent behind the definitions as presented here is to be inclusive of various technologies and design techniques while still drawing distinctions among concepts with different implications for planning practice.

solar access: The ability to receive sunlight across real property for passive solar design or any solar energy system.

solar site design: A passive solar design technique that uses street orientation and lot design to maximize on-site solar energy use potential.

solar collector: Any device that transforms solar radiation into thermal or electrical energy.

solar energy system: A complete assembly consisting of one or more solar collectors and associated mounting hardware or equipment.

solar garden: Any freestanding solar energy system as a primary land use on a site up to 10 acres in size.

solar farm: Any freestanding solar energy system as a primary land use on a site larger than 10 acres.

solar-ready building: Any building specifically designed to accommodate the installation of a solar energy system.

solar development: Any real estate development that includes passive solar design techniques, solar-ready buildings, or solar energy systems.

CHAPTER 3

Visioning and Goal Setting

*Ann Dillemuth, AICP, David Morley, AICP, Erin Musiol, AICP,
Brian Ross, and Chad Laurent*

 The first strategic point of intervention for communities looking to promote solar energy use through planning is visioning and goal setting. Visioning is a participatory planning process that seeks to describe an agreed-upon desired future for a community. To do this, planners actively engage residents and community stakeholders in discussions and exercises in order to identify shared values and aspirations and set goals tied to these values and aspirations. A successful community-based visioning and goal-setting process creates the foundation for all other strategic points of intervention.

Visioning exercises help community members articulate shared values and goals.

Matt Noonkester

When planners engage residents and other community stakeholders in long-range visioning exercises, they help communities determine the values that should undergird their goals, objectives, policies, and actions. Communities that place a high value on harmony with nature and local economic resilience have good reasons to prioritize solar energy use.

This chapter begins with a discussion of why an increasing number of localities are thinking about how their policy frameworks may be supporting or preventing solar energy use. The subsequent sections are intended to provide planners, public officials, and other engaged stakeholders with a primer on some of the key issues that may arise during community conversations about solar development. And the final sections of this primer examine how local visioning and goal-setting processes can both help balance solar energy use with other competing community priorities and set the stage for solar development projects that meet multiple community goals.

SOLAR ENERGY AND THE LOCAL POLICY AGENDA

Why is solar energy part of the local policy discussion? Energy policy was historically the province of regional utilities and state and federal regulators. Utilities built large centralized power plants and transmission lines, and in most states, state and federal law removed the planning decisions from local decision making. Today, however, energy planners recognize that distributed generation will be a significant component of the nation's future energy portfolio. Rather than distant power plants providing a one-way flow of energy to our communities, the future will have both large and small on- and off-grid generation sources. A homeowner or business might be a power or heat generator for part of the day and a consumer for the rest of the day.

Justifications for Prioritizing Solar Energy Use

Local officials have a number of reasons for supporting solar energy use in local policy, plans, programs, processes, and regulations. A local interest in renewable energy issues may be attributable to a growing interest in sustainability or climate action planning. If so, solar development may be viewed as one component of a broader strategy to reduce greenhouse gas emissions or reduce dependence on nonrenewable energy sources. Another component of sustainability is "localization," as seen with the local foods movement, which emphasizes the economic, social, and environmental benefits of local production. Local energy has direct sustainability parallels to local foods, and almost every community has a substantial, but largely unused, local solar resource.

Many local and state economic development advocates also recognize the opportunities associated with solar development. While job and local investment opportunities exist along the entire solar development supply chain, installation jobs and investment are especially noteworthy since they cannot be moved overseas or centralized in a distant community (Figure 3.1).

Local action may also stem from state initiatives. Twenty-nine states have some form of a renewable energy portfolio standard (RPS) for utilities, and 16 have an RPS specifically for solar energy or distributed generation. Additionally, a number of large and small municipal utilities not covered by a state policy have set their own equivalent renewable procurement targets. An RPS requires utilities and energy suppliers to obtain a certain amount of the energy they use to serve their customers from renewable energy resources. While the policies are enacted at the state level, local governments have opportunities to support implementation. As every community has solar resources, each community needs to consider how solar development occurs in its jurisdiction.

Moreover, as solar development becomes more commonplace, conflicts and concerns about how and where development occurs are likely to become more prevalent. Rather than wait until a conflict arises, local officials should lay the foundation for decision making on individual projects that will require balancing of different community goals. For example, solar development may conflict in some instances with tree protection and urban forestry goals, historic and heritage preservation goals, and community character standards. Similarly, policies and actions that support these other goals may limit opportunities to use solar energy (see Competing Priorities below).

Characteristics of Solar-Friendly Communities

Several states have developed programmatic approaches to help communities understand how to implement solar-friendly measures. For example, the Colorado Solar Energy Industry Association developed a solar-friendly community certification program that has also been used in other states. California completed a detailed guideline for communities to implement permitting best practices within the context of California statutes and policies. Minnesota developed model templates for plan policies, development regulation, and building permit applications designed to integrate with local processes typical for Minnesota. But in all cases, solar-friendly communities have some common characteristics.

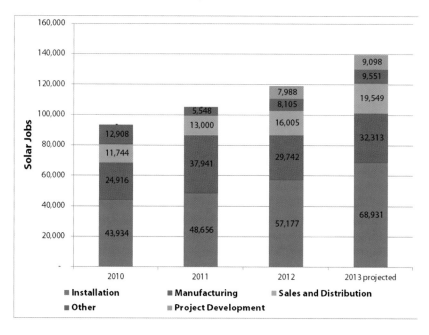

Figure 3.1. Employment in installation and sales of solar energy systems has grown steadily since 2010.

Source: Solar Foundation et al. 2012

These characteristics include:

- comprehensive plans and other policy documents that acknowledge the community's solar resources as a valuable asset;

- development regulations that clearly identify as-of-right solar installation opportunities for different types of installations, clear requirements and reasonable processes for installations that are not as-of-right, and a means of protection of long-term access to direct sunlight for energy production;

- permitting and inspection processes that are transparent, predictable, and easily accessible for contractors to use in preparing bids and counter staff and inspectors to use to ensure a consistent review and inspection process;

- an integrated process of approvals with the electric utility for interconnecting solar developments to the grid; and

- public-sector investment in solar resources to demonstrate both feasibility and community commitment to using local resources.

Appendix A includes a checklist with specific questions for communities to consider as they evaluate how well their plans, development regulations, and permitting processes support solar energy use.

COMMUNITY ENGAGEMENT

Planners have an important role to play in initiating and facilitating community conversations about solar energy. These conversations may be in the context of formal visioning or goal-setting exercises; alternately, questions or concerns about solar energy may rise spontaneously in response to specific development proposals.

Through visioning and goal-setting exercises, planners have opportunities to initiate community conversations about solar energy, and these exercises give planners opportunities to highlight both the benefits of and barriers to increasing local solar energy production. While visioning is an ideal venue for initiating a community conversation about solar energy, planners should also be prepared to facilitate conversations about solar that might arise either in response to a specific development proposal or through some other phase of the planning process. Discussions of policy or project alternatives may segue naturally into a community conversation about solar. When this happens, planners must be prepared to provide complete and accurate information about solar energy and how it connects to other community goals and values.

Addressing Concerns

In a 2011 survey, local governments identified a lack of understanding of solar technology as one of the top five challenges to expanding solar energy use (ICMA 2011). Planners can play an important role in addressing this challenge by providing accurate information about solar energy use, regardless of the forum. The following sections provide facts in order to help address some of the most common concerns related to solar energy systems; it is important to note that passive solar design seldom raises similar concerns.

Concerns about the adequacy of the solar resource. Planners should understand that, though individual sites vary in the amount of solar insolation they receive due to latitude, atmospheric conditions, and the local landscape, on the whole their communities receive enough annual solar radiation to

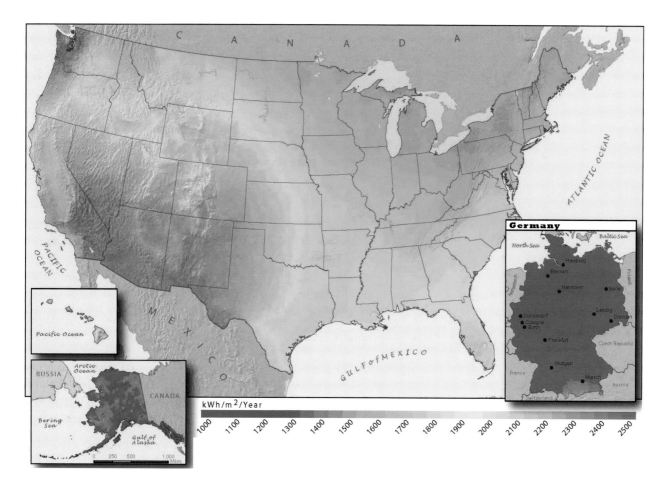

make solar energy use a viable option. Every state in the U.S. receives as much, or more, sunlight than Germany, which leads the world in solar PV installation and energy production (Figure 3.2).

Concerns about system costs. While solar energy systems do require up-front capital investments, the cost of solar PV has been dropping rapidly over the last few years due to improvements in manufacturing processes and economies of scale. A range of financial incentives—including grants, rebates, low-interest loans, and tax credits—are available from federal, state, and local governments, as well as utilities, to further offset the initial up-front costs of PV and solar thermal systems (DSIRE 2013b). Prices of PV equipment will most likely continue to fall and efficiencies will continue to increase, but if solar makes economic sense today, there is no reason to wait to install a system.

Solar PV and thermal systems are often sound investments; homeowners benefit from reduced energy bills and savings many years after their systems have paid for themselves, and studies have shown that solar energy systems add value to homes (Farhar and Coburn 2006; Hoen et al. 2011). Local governments across the country are adding solar energy systems to city buildings, parking lots, and other structures to help reduce their energy bills for building and plant operations over the long term. In addition, third-party financing arrangements, including leases and power purchase agreements (PPAs) as well as cooperative solar projects in which consumers purchase shares in a centralized solar energy facility, are opening up the benefits of solar energy to those who may not own property or who lack the funds to purchase their own energy systems (Wesoff 2012).

Concerns about economic viability. Though much is made of the financial incentives allotted for renewable energy production at both the federal and state levels, these are a drop in the bucket compared to the subsidies that

Figure 3.2. *Most areas of the U.S. receive more insolation than Germany, the current world leader in solar power generation.*

National Renewable Energy Laboratory

Note: Annual average solar resource data are for a solar collector oriented toward the south at a tilt = local latitude. The data for Hawaii and the 48 contiguous states are derived from a model developed at SUNY/Albany using geostationary weather satellite data for the period 1998–2005. The data for Alaska are derived from a 40-km satellite and surface cloud cover database for the period 1985–1991 (NREL 2003). The data for Germany were acquired from the Joint Research Centre of the European Commission and are the yearly sums of global irradiation on an optimally-inclined surface for the period 1981–1990.

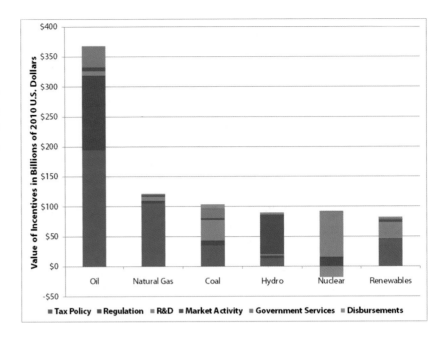

Figure 3.3. *Summary of federal energy incentives, 1950-2010 (billions of 2010 dollars)*

Source: MISI 2011

the fossil fuel industry has received for nearly 100 years (Pfund and Healey 2011) (Figure 3.3). Further, oil and gas subsidies are stable compared to the short-term and unpredictable incentive landscape for solar energy. Yet the cost per watt of this energy source continues to drop despite these comparative shortcomings in economic support, and solar is expected to equal the cost of other electricity sources even without subsidies by 2020. Likewise, statistics show solar to be a robust and growing industry, with continuing expansion expected in both future U.S. manufacturing capacity and solar installations (SEIA and GTM Research 2013).

Concerns about environmental impacts. Though solar technology and manufacturing may be complex, solar collector composition is fairly simple. Most PV panels are constructed of glass (silicon), with common metals such as aluminum and copper wiring. With the exception of thin-film solar products, which may contain heavy metals, PV panels do not contain potentially toxic substances. More than 90 percent of a PV module can be recycled at the end of its productive life. Furthermore, a number of manufacturers offer voluntary panel take-back programs (Sniderman 2012). Similarly, solar thermal

Figure 3.4. *Percentage of sunlight reflected by different surfaces at different angles of incidence*

Source: Shields 2010

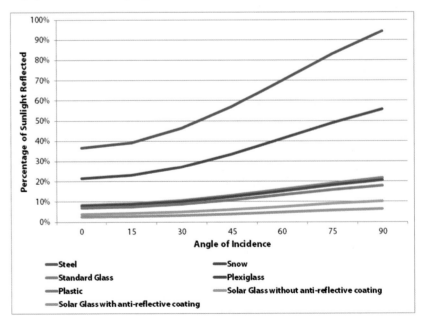

collectors consist largely of copper, aluminum, steel, or polymers, most of which are easily recyclable and nontoxic. Like all manufactured products, the production of solar systems does cost energy—however, studies show that the energy produced by solar PV and thermal systems more than pays off the energy cost of their manufacture, often with energy-cost paybacks of two years or less (Fthenakis 2013; Streicher et al. 2004).

Concerns about glare. Some residents may express concerns that glare from solar collectors will be either a public or private nuisance. However, because they are constructed of dark-colored materials and covered with anti-reflective coatings, new solar PV and thermal systems typically reflect as little as 2 percent of incoming sunlight (Figure 3.4). In fact, a number of solar installations have been successfully located at or near several U.S. airports (including Boston, New York, San Francisco, and Denver), where glare is of paramount concern, and evidence thus far suggests that glare has not been a problem for airport personnel at these locations (FAA 2010). That said, solar collectors are made of smooth glass that is capable, under some circumstances, of producing a concentrated reflection, and the potential for glare is greater when the sun is low on the horizon. As a consequence, Sandia National Laboratories (2013) has developed a series of tools that allow users to assess glare potential or effects.

Highlighting Opportunities

Beyond addressing any concerns about solar energy that constituents may have, planners can help raise local awareness about solar energy use opportunities. There is a need for public education and outreach around solar energy; "lack of interest in or awareness of solar energy development" was the third most commonly reported challenge in the 2011 solar survey (ICMA 2011). Planners can assist with local solar efforts by developing materials about and conduits for information on solar technology, policies, and programs to help educate the public, as well as facilitating hands-on opportunities for residents to learn more about solar energy use.

In some cases, planners and other local officials may not have the expertise, the time, or the resources to provide solar energy information or education. Consequently, many cities and counties have developed relationships with local solar experts to help promote awareness in their communities. This can include hiring consultants or partnering with nonprofits, local industry professionals, or educators to provide information or educational opportunities related to solar energy use.

COMPETING PRIORITIES

Most communities pursue multiple goals simultaneously through a range of plans, policies, regulations, and programs. The decisions communities make in support of one goal may have a positive, negative, or negligible effect on other goals. When a community considers each goal in isolation, it may miss opportunities to address potential conflicts before they occur. Once a conflict exists, it may be too late to pursue a mutually beneficial solution, and communities may be forced to choose between competing interests.

As Godschalk and others have pointed out, sustainability goals are not immune to these potential conflicts of interest (Godschalk 2004; Campbell 1996). When the goal is promoting the installation of solar energy systems, other goals such as tree protection, historic preservation, and even urban redevelopment may represent competing interests. Fortunately, planners' comprehensive approaches to problems and long-range perspectives allow them to consider potential tradeoffs and find ways to balance different—and sometimes competing—community priorities and goals. Moving forward, planners can serve as key players in ensuring that solar development can coexist with other potentially competing interests.

EXAMPLES OF SOLAR COMMUNITY OUTREACH ACTIVITIES

- Seattle developed a guidebook on solar energy system permitting (www.seattle.gov/DPD/Publications/CAM/cam420.pdf) describing solar PV and hot water systems and outlining permitting and land-use requirements, design and installation standards, contractor selection considerations, and financial incentives.

- San Francisco maintains a solar map (http://sfenergymap.org/) highlighting existing solar PV and water heating installations in the city (identifying locations, system sizes, and installers) and allows users to enter a city address to find a property's solar-electric and water-heating potential.

- Knoxville, Tennessee, hosts a solar energy website (www.solarknoxville.org) providing basic information about solar PV and water heating systems, links to local installers, a list of exemplary local solar installations, and details about local events, workshops, and solar tours.

- Santa Barbara, California, offers a solar recognition program with annual awards for solar projects completed in compliance with the city's Solar Energy Systems Design Guidelines.

- Austin, Texas, worked with science coordinators and curriculum directors from the local school district to develop hands-on learning curriculum materials designed to be used in conjunction with 13 solar installations at local schools (USDOE 2011a).

- Madison, Wisconsin, hired a consultant to act as a "solar agent" for home and business owners, performing free site surveys for interested residents and helping arrange onsite and financial assessments of potential systems (USDOE 2011b).

Tree Protection

Maintaining and enhancing the tree canopy is a common sustainability goal. Trees provide a wide range of environmental, social, and economic benefits, including improving air quality, reducing stress, and increasing property values. When a tree's shade impacts the efficiency of a solar system, however, trees and solar become unlikely adversaries. The conflict has sparked debates about which should be the higher local priority.

Despite the many benefits of trees, urban tree coverage is on the decline across the U.S. (Nowak and Greenfield 2012). Solar energy systems could represent another potential threat to an already increasingly threatened resource. Some states require the removal of trees that could grow and interfere with solar energy systems, even if the trees were planted prior to the installation of the system, and alternatives to tree removal such as pruning and height restrictions can reduce the benefits of the tree canopy. Areas with high concentrations of solar energy systems may effectively become buffers against future tree plantings. Tree advocates worry about the implications of today's solar installations on the future of the urban forest.

Several types of disputes related to trees and solar energy systems are common at the local level. One relates to property owners who would like to cut down trees on their properties to install solar systems but are prohibited from or charged fees for doing so by local regulations. Another arises when a neighbor of a property with a solar energy system already in place plants trees that are likely to grow to block solar access. A third type occurs when existing trees grow to block a new solar installation on a neighboring property. Trees often lose when solar energy systems and trees conflict, even if they were planted before the solar energy systems were installed.

In the United States, there is no common law "right to light," and there is no federal statute or policy addressing or affirming solar rights (Staley 2012a). The patchwork of solar access regulations at the state level is reflected at the local level (see Chapter 5). Many communities have developed plans and regulations to protect their urban forests, but only a few address potential conflicts between trees and solar installations. One example is Ashland, Oregon, which offers a solar access permit that protects against shading of solar energy systems by vegetation (§18.70). Once a property owner has recorded a system and obtained a permit, no vegetation may be allowed to shade the system, and the city has the power to declare shading vegetation to be a nuisance and enforce mandatory tree trimming. Vegetation over 15 feet in height at the time the permit is applied for is exempt, which helps protect existing trees.

The variety of regulations, guidelines, and policies that exists pertaining to solar and trees reveals the challenges communities face in trying to prioritize these two valuable resources. Instead of outcomes where one resource wins out over the other, communities should refocus their efforts on taking measures to ensure these interests can successfully coexist. The following recommendations can assist planners in these efforts:

• Ensure that the right tree is planted in the right place and for the right reason to minimize the chance of conflict at a later date (Staley 2012b).

• Address urban forests and solar together during the comprehensive planning process to provide a basis for addressing this issue in ordinances, development review, and code enforcement (Staley 2012b).

• Consider creating and adopting overlay zoning for "solar access zones" in suitable areas that specifically acknowledges the need to consider plant size to maintain clearance for solar collection (Staley 2012b).

- Invite and encourage urban foresters to become members of local solar advisory committees and councils.

- Replace removed trees where possible, and track tree removals to ensure there is no net loss in trees.

- In instances where a solar installation would result in the removal of mature trees, encourage or require other energy conservation strategies first along with pruning of trees rather than tree removal.

- Actively identify the best places to locate solar energy systems in a community—developed areas where infrastructure is already in place, such as parking lots, roads, brownfield and greyfield sites, landfills, and big-box stores—and direct installations to these areas.

- Educate citizens as to the benefits of both solar and trees, and increase their awareness of best practices of sensible planning to avoid shading and ensure that solar and trees can coexist.

Historic Preservation

Both historic preservation and solar energy use are often included in community plans to increase sustainability. Like solar energy, historic preservation is both environmentally friendly and economically beneficial. Historic properties were typically built with attention to climate and air circulation and with locally sourced materials, and preservation of historic properties is "greener" than tearing down and rebuilding because of the energy and materials savings (WBDB 2012). Designating a property or district as historic increases property values and attracts investment in and around the area (NTHP 2011). However, tension has developed between these two interests as communities struggle with how to both preserve their pasts and ensure sustainable futures.

Changes to a building's structure or façade to support a solar installation, as well as improper placement of an installation, can threaten the historic character and architectural integrity of historic resources. Neighboring property owners and other stakeholders may fear related reductions in property values. Proponents of solar, however, feel that solar technology can help

A PV system was installed on the rear elevation of this historic property in the Heritage Hill Historic District of Grand Rapids, Michigan. By locating the system in the rear of the property, the views from the public right-of-way remain preserved.

Kimberley Kooles

strengthen the environmental profile of older buildings and help jurisdictions meet aggressive energy goals (Musser 2010). How a community chooses to address this potential conflict can greatly impact its ability to maximize its solar potential or to protect its historic resources.

As noted above, many states have enacted solar rights legislation. Some states allow "reasonable" restrictions, though this is defined differently from state to state and in most cases it is unclear whether historic preservation falls into this category (Kettles 2008). A handful of states have specifically addressed the issue: North Carolina authorizes local jurisdictions to regulate the location or screening of solar collectors in historic districts and more generally when systems would be visible from public rights-of-way (N.C. Gen. Stat. §160A); New Mexico prohibits a county or municipality from imposing restrictions on the installation of solar collectors except in a historic district (N.M. Stat. §3-18-32); and Connecticut prohibits preservation commissions from denying certificates of appropriateness for renewable energy systems in historic districts unless the system "substantially impairs" historic character, though they may impose design and location conditions (Conn. Gen. Stat. § 7-147f).

Some local jurisdictions have taken steps to explicitly address solar and historic preservation in their codes and ordinances. For example, Howard County, Maryland, and Alexandria, Virginia, have adopted guidelines for solar panels in historic districts (Howard 2009; Alexandria 1993). The zoning ordinance of Austin, Texas, allows preservation plans in historic districts to incorporate sustainability measures such as solar technologies and other energy generation and efficiency mechanisms (§25-2-356; §25-2-531). Montgomery County, Maryland, amended its General Rehabilitation Design Guidelines in 2011 to specifically address solar panels (MCPD 2011). Portland, Oregon, revamped its zoning code to eliminate discretionary review of all new solar installations that comply with community design standards (Chap. 33.218).

However, these localities are the exceptions rather than the rule. When communities do not address solar energy in plans, ordinances, or design guidelines, the resulting uncertainties can discourage the installation of solar systems or cause property owners unaware of requirements to install systems without required approvals. Some communities may allow solar in historic districts or on historic properties, but other provisions in their ordinances impose so many obstacles and restrictions on permit approvals that installing solar energy systems becomes unfeasible or impossible. Finally, solar installations in historic contexts are often considered on a case-by-case basis, leaving municipal review boards, commissions, and councils to resolve solar and historic preservation conflicts through discretionary decisions rather than clear standards.

Though there is currently no standard approach to determining whether a solar installation is appropriate on a historic resource, most agree that solar is not acceptable when the installation involves removal of historic roofing materials, when the historic roof configuration has to be removed or altered, or when the installation procedure would cause irreversible changes to historic features. Panels are generally viewed as acceptable when they are installed on flat roofs and are not visible from the street; installed on secondary facades and shielded from view from a primary façade; ground-mounted in inconspicuous locations; located on new buildings on historic sites or new additions to historic buildings; and are complementary to the surrounding features of the historic resource (Kooles et al. 2012).

The following recommendations will assist planners in ensuring that solar energy systems and historic preservation can successfully coexist in their communities:

- Address historic preservation and solar energy systems jointly throughout the planning process by discussing priorities and potential conflicts.

- Revise, develop, and adopt local preservation guidelines or ordinances (tailored to the community) that address renewable energy and sustainable technology.

- Perform an audit of the community's historic preservation guidelines and regulations to determine unnecessary barriers to solar installations.

- Ensure that both historic preservationists and solar experts are involved in the development of solar access guidelines and development regulations and serve as members of local solar-advisory committees.

- Designate a board with appropriate stakeholder representation to make decisions regarding solar energy systems and historic structures.

- Educate and increase citizen awareness of the benefits of both solar energy systems and historic preservation and best practices of sensible planning to avoid future conflicts.

Urban Redevelopment

A potential competing interest with solar energy systems that remains largely overlooked is urban redevelopment. Many communities seek to concentrate development in targeted areas like downtowns or transit-oriented developments in an attempt to reduce vehicle miles traveled, provide more efficient services, and offer transportation and housing alternatives. This often means changes to regulations, including height restrictions, to accommodate future growth. At the same time, these are the same areas where communities are often encouraging solar projects. Just as shade cast by a tree over a solar energy system can reduce the installation's efficiency, so can the shadow of a tall building. As targeted areas redevelop, the possibility for solar conflicts rises.

Though many states have adopted legislation in attempts to ensure that existing solar installations have access to an adequate amount of sunlight, even the states with notable solar rights legislation have not specifically addressed the issue of solar and urban redevelopment. To date there have

Shadows from new tall buildings have the potential to shade existing solar development.

Source: Erley and Jaffe 1979

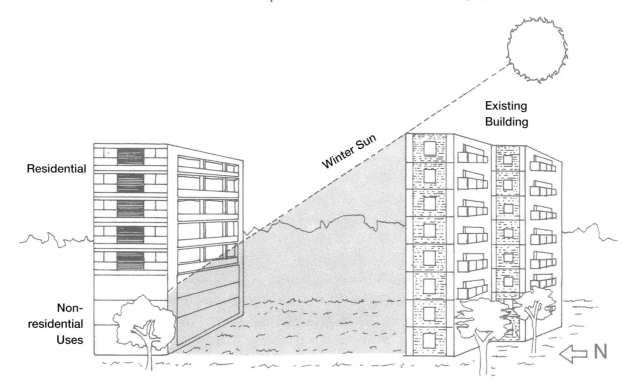

been little to no documented disputes in the U.S, but the potential for future conflict is high.

Communities need to think through the potential tradeoffs and develop strategies to address conflicts before they arise. This may involve determining prime areas for both solar installations and redevelopment and identifying them for the public through the use of tools like overlay districts. When these areas overlap, communities could develop design guidelines or standards that specifically address the impacts of massing on existing solar installations. Additionally, communities that have adopted solar-access zone provisions mitigating conflicts with trees can modify those provisions to address redevelopment as well. With no solar access laws in the U.S. specifically addressing this issue, communities should also consider looking internationally to see how other countries' solar access laws for urban areas are evolving.

VACANT LAND MANAGEMENT AND SOLAR DEVELOPMENT

In addition to helping to minimize potential conflicts associated with solar energy use, local visioning and goal-setting processes are also beneficial in identifying synergies between solar energy use and other community priorities. Perhaps the most important such potential synergy is the opportunity to meet multiple community goals by recycling underutilized or surplus vacant land for solar development.

In contrast to sites prioritized for dense redevelopment, primary-use renewable energy facilities can be a great match for brownfields or other vacant sites in weaker market areas. The site cleanup requirements for a solar farm or garden are typically less extensive and costly than they would be for recreational, commercial, or residential uses, and these installations can be dismantled and moved to make way for a higher and better use of the property if other redevelopment opportunities eventually arise.

Communities plagued with high numbers of vacant properties face serious headwinds for economic recovery. In some areas, former industrial sites pose the biggest challenges, while in others the most obvious signs of distress may be vacant houses or shuttered strip retail. While many of these properties will eventually find new life through re-occupancy or conventional redevelopment, weak market areas may need to embrace alternative reuse options like urban agriculture, open space, or renewable energy production.

Recycling weak-market vacant land for solar energy projects is worth prioritizing for numerous reasons. Construction and installation work creates demand for local green-collar jobs, and reactivating a vacant site with a solar installation can reduce blight and improve appearances. When a

The 1.3 MW PSE&G Trenton Solar Farm, located on a 5.5 acre remediated brownfield, provides enough electricity to power about 200 homes annually.

PSE&G

solar redevelopment project involves cleanup of a contaminated site, it has the dual benefit of decreasing public health risks and repairing damage to the natural environment. Furthermore, redevelopment projects provide an alternative to developing on greenfields, and previously developed sites are typically well-positioned to take advantage of existing infrastructure and public services.

While it is true that freestanding and rooftop solar energy systems can be a good fit for vacant properties of all sizes in a wide range of contexts, there can still be serious barriers to recycling land for solar energy production. These barriers may include incomplete or inaccurate information about available sites, inadequate solar access, outdated or confusing development regulations, extensive on-site contamination, and insufficient project financing.

SUMMARY

Visioning and goal setting is the first strategic point of intervention for communities looking to promote solar energy use through planning. Long-range visioning and goal-setting exercises help communities determine the values that should undergird their local policies, plans, programs, processes, and regulations. Planners, local officials, and other community stakeholders typically become interested in supporting solar energy use for environmental or economic reasons. However, because most communities pursue multiple goals simultaneously, it is important for planners to help community stakeholders understand how solar energy use affects other community resources. In some cases the task is primarily one of providing accurate information to dispel misconceptions or highlight existing opportunities. In other cases, though, visioning and goal-setting exercises can help identify both potential conflicts and synergies between solar energy use and other community priorities.

CHAPTER 4

Plan Making

David Morley, AICP, and Erin Musiol, AICP

 The second strategic point of intervention for communities looking to promote solar energy use through planning is plan making. Communities adopt local plans in order to chart courses for more sustainable and livable futures. Planners and public officials then use these plans to inform decisions that affect the social, economic, and physical growth and change of their communities. Given the potential economic and environmental benefits of local solar development, it is no wonder that an increasing number of cities and counties are addressing solar energy use in their plans.

American Planning Association

Figure 4.1. *There are at least three distinct types of local plans: communitywide comprehensive plans, subarea plans, and functional plans.*

Most local plans fall into one of three broad categories: (1) comprehensive plans, (2) subarea plans, or (3) functional plans (Figure 4.1). Comprehensive plans cover a wide range of topics of communitywide importance. In contrast, subarea plans cover one or more topics of particular importance to a limited part of a single jurisdiction, and functional plans focus on a single topic or system that is not limited to a single subarea. While some communities have adopted functional plans on the specific topic of solar energy use, many others address solar energy in comprehensive or subarea plans or in functional plans covering climate change, sustainability, or energy.

This chapter explores how communities can incorporate solar-supportive commentary and goals, objectives, policies, and actions into various types of local plans. The first section summarizes the roles that four common plan features can play in solar energy use. The remaining sections take a closer look at specific types of plans.

COMMON FEATURES OF LOCAL PLANS

While local plans vary based on geographic scale, timeframe, and breadth of topics, there are four features common to most local plans: (1) an explanation of the purpose of the plan, (2) a discussion of existing conditions and trends, (3) a presentation of desired outcomes in the form of goals and objectives, and (4) an enumeration of policies and actions in support of these goals and objectives. Planners and others involved in plan making have opportunities to address solar energy use in each of these common plan sections.

Plan Purpose

The purpose section allows plan authors to explain the plan's impetus, scope, and authority, while also providing some insight into the nature of the planning process. In practice, the plan purpose may be articulated in one or more introductory paragraphs, or it may be stated broadly up front and then revisited or reframed in the introduction to each thematic plan element.

Regardless of the form of the statement in a particular plan, a core purpose of local planning is facilitating the development or protection of community resources. Because sunlight can be harvested for heat or electricity, it has a value beyond its intrinsic human health benefits. However, relatively few communities acknowledge solar energy as a resource comparable to other local resources such as vegetation, water, minerals, fossil fuel reserves, or historical buildings and heritage sites. Certain Minnesota communities are the exception, as the Metropolitan Land Planning Act governing the Twin Cities metropolitan area requires all constituent jurisdictions to address solar access in their comprehensive plans. Victoria, Minnesota, provides an example: the Special Resources section of its comprehensive plan identifies solar energy as having the greatest potential to replace limited fossil fuel supplies and provides goals and policies for the protection of solar access and for alternative energy development (Victoria 2009).

Unless the scope of the plan is limited to identifying opportunities to promote solar energy use, a plan purpose statement may not explicitly reference solar energy. However, one way to help community members start to see solar energy as a local resource is by simply pointing out the nonlocal and nonrenewable origins of most locally used energy. For example, the Solar Access Protection element of Shakopee, Minnesota's 2030 Comprehensive Plan begins with a warning that the State of Minnesota currently produces only 0.2 percent of the fuel it uses (Shakopee 2009).

Existing Conditions and Trends

The existing conditions and trends section provides context for the broad goals and objectives of the plan and sets the stage for the policies or policy considerations detailed in subsequent plan sections. Understanding the potential importance of a community's solar resource requires some knowledge of both the availability of the local solar resource and the community's existing energy use. Plan authors can use this section to document the amount of energy consumed, the mix of energy sources currently used in the community, information about the local prevalence of passive solar design techniques, existing installed solar capacity, and a summary of how local solar investment has changed over time.

While a national map of insolation levels in the U.S. will show that solar energy production is a viable option across the country, demonstrating how access to sunlight may vary across a community can be especially helpful. The potential for harvesting this resource in a specific location depends primarily on local landscape variables rather than general conditions such as latitude and average cloud cover. Topography and shading from trees and adjacent buildings have a greater impact on the available solar resource for a specific building or site than whether or not the property is located in Arizona, Minnesota, or North Carolina.

Identifying the areas with the greatest potential for solar energy use in a community can help residents, business owners, and developers understand how to direct their efforts and investments, and from a broader perspective, this information helps elected officials make decisions about where to focus future development and conservation efforts. Many communities have developed solar maps to assist in this process (see Chapter 7). These interactive online tools allow users to calculate available solar radiation at the parcel level and often provide estimates of the potential energy generation and gas or electric bill savings that could be realized through solar energy system installation on a specific property. Solar maps are also used to track existing solar installations and total installed capacity within communities.

Once a community has documented the geographic characteristics of its solar potential, plan authors can summarize this information for community members. For example, the latest version of Los Angeles County's Conservation and Natural Resources General Plan Element includes a direct link to the county's online Solar Map and Green Planning Tool, and Nye County, Nevada, includes maps indicating solar suitability and documenting existing solar installations as figures in the conservation chapter of its 2011 Comprehensive Master Plan.

The existing conditions and trends section can also describe how technological and economic factors influence local solar energy use. As stated in Chapter 2, different solar technologies are associated with different costs and performance characteristics. Furthermore, the feasibility of different types of solar energy systems is influenced by state and local policies and incentives. While an inventory of state and local policy incentives for solar energy use may be too detailed for an existing conditions analysis in a comprehensive or subarea plan dealing with multiple topics, a functional plan focused exclusively on energy may include a summary of the state and local laws that encourage or inhibit solar energy use.

Beyond the state and local policy context, financial incentives can also exert a powerful influence on the local solar energy market. Unlike regulations, which tend to be relatively static, incentives are often more ephemeral. Consequently, instead of cataloguing existing state and local incentives for solar installations, it may make more sense for plan authors to describe the different species of incentives that have historically influenced local solar markets. For example, some common types of financial incentives for solar installations are renewable energy credits, tax credits, and equipment rebates. Plan authors can then reference an up-to-date resource, such as the Database of State Incentives for Renewable Energy (DSIRE), to direct readers to more information about active programs and offers.

Perhaps the most easily overlooked factor influencing the local solar market is the knowledge and experience of local solar installers. Like incentives, local installer capacity is dynamic and can change rapidly as new installers enter the marketplace or existing installers acquire additional experience or training. While plan authors may not be able to discuss local capacity in detail, they can help community members understand how limited installer knowledge and experience may present a barrier to expanding the local solar market.

One example of a community that has included extensive information about baseline conditions and trends related to solar energy is Tucson, Arizona. The city's Solar Integration Plan and its companion, the Greater Tucson Solar Development Plan: Strategies for Sustainable Solar Power Development in the Tucson Region, provide a thorough analysis of existing solar capacity and the factors likely to influence market growth.

Goals and Objectives

Almost all local plans contain one or more sections presenting goals and objectives related to the plan's focus. Goals are general statements about desirable future conditions. Objectives are statements of measurable outcomes in furtherance of a certain goal. Together, goals and objectives make up the cornerstones of a local policy framework. In other words, all local policies and implementation actions should ideally be in furtherance of adopted goals and objectives.

Most communities formulate and prioritize goals through participatory planning processes. These processes may include formal visioning and goal-setting exercises as well as various other citizen engagement tools used by planners to facilitate conversations about the future of their communities. Planners have important roles to play in engaging the community in developing goals related to solar energy.

Each goal in a plan may be associated with multiple objectives. Again, these objectives are ideally byproducts of robust and authentic participatory planning. Effective objectives are both achievable and subject to measurement. In order to formulate an achievable objective, participants in the planning process must have access to the best available data and analyses on the issues at hand. Moreover, objectives should, ideally, be associated with a timeframe. This may be the time horizon of the overall plan, or it may be specific to the objective.

The process of formulating goals and objectives may be the first and best opportunity for planners and other participants to work through potential conflicts among goals. There are innumerable actions that communities can pursue to improve sustainability and livability for their residents; however, not all of these actions are necessarily compatible with each other, and in the case of solar energy use, other sustainability goals such as tree preservation, historic preservation, and urban redevelopment can create potential conflicts. Addressing solar energy use in the context of these other areas early in the planning process can help communities develop policies and regulations that can promote successful outcomes if conflicts arise. See Chapter 3 for further discussion of potentially competing goals.

Among communities that have added goals and objectives to their plans related to renewable energy, generally, or solar energy, specifically, common themes include encouraging solar site design for new subdivisions, improving the energy performance of municipal facilities, removing barriers and creating incentives for small-scale installations, and capturing economic development opportunities associated with renewable energy investment.

Policies and Actions

Effective local plans typically include both specific policy statements and action steps. Policies are statements of intent with enough clarity to guide decision making, and actions are directives about programs, regulations, operational procedures, or public investments intended to guide the implementation of specific policies.

While goals and objectives generally remain abstract, policies point to a course of action and suggest responsibility for implementation. To illustrate, a goal to expand local renewable energy production, with an objective of increasing installed solar capacity to a certain level by a target date, says little about what roles local government, private developers, and individual property owners should play in order to meet this goal. However, a policy stating that

EXAMPLES OF SOLAR-SUPPORTIVE LOCAL PLANS

The following comprehensive, subarea, and functional plans mentioned in this chapter include commentary or goals, objectives, policies, and actions that support solar energy use:

- Amherst, Massachusetts. Atkins Corner Sustainable Development Plan (2002), Chapter 5, Architecture, Uses and Programming (http://amherstma.gov/DocumentCenter/Home/View/380).

- Anaheim, California. General Plan (2004), Green Element (www.anaheim.net/generalplan/).

- Austin, Texas. Brentwood/Highland Combined Neighborhood Plan (2004), Greenbuilding and Sustainability (ftp://ftp.ci.austin.tx.us/npzd/Austingo/brent-highland-np.pdf).

- Chico, California. 2030 General Plan (2011), Chapter 2, Sustainability; Chapter 3, Land Use; Chapter 8, Housing Element (www.chico.ca.us/document_library/general_plan/GeneralPlan.asp).

- City of Fort Collins, Colorado. City Plan (2011), Environmental Health Principles and Policies – Energy (www.fcgov.com/planfortcollins/pdf/cityplan.pdf).

- Lawrence Township, New Jersey. The Green Buildings and Environmental Sustainability Element of the Master Plan (2010), Energy Conservation and Renewable Energy Production (www.lawrencetwp.com/documents/planning/Lawrence%20Sustainability%20Element.pdf).

- Los Angeles County, California. General Plan 2035 Public Review Draft (2012), Chapter 6, Conservation and Natural Resources Element (http://planning.lacounty.gov/generalplan/draft2012).

- Nye County, Nevada. 2011 Comprehensive/Master Plan (2011), Section 3, Conservation Plan – Renewable Energy – Solar, Geothermal, Wind and Biomass (www.nyecounty.net/DocumentCenter/Home/View/14049).

- Orlando, Florida. Growth Management Plan (2010), Conservation Element (www.cityoforlando.net/planning/cityplanning/GMP.htm).

- Pinal County, Arizona. Comprehensive Plan (2009), Chapter 7, Environmental Stewardship (http://pinalcountyaz.gov/Departments/PlanningDevelopment/ComprehensivePlanUpdate/Documents/00.Comprehensive%20Plan%202012.pdf).

- Shakopee, Minnesota. 2030 Comprehensive Plan Update (2009), Solar Access (www.ci.shakopee.mn.us/compplanupdate.cfm).

- Tucson, Arizona. Solar Integration Plan (2009) (http://cms3.tucsonaz.gov/files/energy/Solar%20Plan%20Final.pdf).

- Tuscon, Arizona. Great Tucson Solar Development Plan: Strategies for Sustainable Solar Power (2009) (http://www.pagnet.org/documents/solar/SolarDevPlan2009-01.pdf).

- Victoria, Minnesota. 2030 Comprehensive Plan Update (2009), Section 2(L)(1), Solar Access Protection (www.ci.victoria.mn.us/documents/2030ComprehensivePlan-Final.pdf).

rooftop residential solar installations should be permitted in all areas of the jurisdiction implies that the local legislative body will adopt new zoning regulations for accessory solar energy systems.

Action steps make the implied responsibilities of policy statements explicit. For example, a plan with a policy sanctioning residential solar installations may include a directive for the planning staff to prepare a zoning amendment for city council review that defines accessory solar energy systems and permits these systems by right in all districts.

Among communities that have included policies and actions related to solar energy use in their plans, common topics include the addition of solar energy systems to municipal facilities, solar access protection, regulatory or financial incentives for small-scale solar installations, and preferential locations for new solar energy systems. Fort Collins, Colorado, and Chico, California, are two examples of communities that make clear connections among solar-related goals, policies, and actions in their most recent comprehensive plans. Both plans include multiple policies aimed at expanding passive solar design and solar energy systems on public and private properties, and both plans detail specific actions and identify parties responsible for implementing these policies.

SOLAR IN THE COMPREHENSIVE PLAN

While planners help towns, cities, and counties prepare a wide range of communitywide, subarea, and functional plans, the most significant of these is the local comprehensive plan. The comprehensive plan, sometimes referred to as the general plan or the master plan, is the foundational policy document for local governments. In many ways it functions like a community constitution, establishing a framework for future growth and change within the jurisdiction to be implemented through local laws and public investments over the next 20–25 years. Given the importance of the comprehensive plan in the local planning system, it represents a logical point to introduce goals and objectives related to solar energy use in the context of the wider local policy framework. This gives plan authors a chance to highlight synergies and potential conflicts between solar development and other community resources and to summarize any previous, ongoing, and planned policies and actions to support the implementation of goals related to promoting solar energy use.

Comprehensive plans are named as such because they cover a broad range of topics of communitywide concern. All states either allow or require local governments to prepare comprehensive plans, and many states require local development regulations to be in conformance with an adopted comprehensive plan. While enabling laws vary from state to state, common topics for plan elements (i.e., chapters or major sections) include land use, transportation, housing, economic development, and community facilities. In recent years an increasing number of cities and counties have added elements addressing sustainability, natural resources, or energy to their comprehensive plans.

The comprehensive plan is the legal foundation that legitimizes local land-use regulations. As such, it is important for plan authors to establish a policy foundation in the comprehensive plan for development regulations that affect solar energy use. Ideally, the local comprehensive plan is a primary guide not only for updates to development regulations but also for the creation of local capital improvements plans, which detail planned capital expenditures over a multiyear period. By extension, comprehensive plans with goals, objectives, policies, and actions that support solar development can pave the way for future public facility construction or rehabilitation and private development projects that incorporate passive solar design or solar energy systems.

In the most recent versions of their comprehensive plans, Orlando, Florida, and Anaheim, California, both tie previous and ongoing activities that support solar energy use to new policies that support local solar market growth. Orlando's plan voices support for ongoing partnerships with the local utility commission and county government to support renewable energy initiatives and includes policies calling for the creation of a solar mapping tool and a solar master plan. Meanwhile, Anaheim's plan references an existing city-owned solar installation and discusses ongoing public education efforts before listing policies that clarify the city's intent to support solar energy systems and passive solar design in both new and existing development.

The first section of Appendix A in this report contains a series of questions for community stakeholders to consider when updating a comprehensive plan. Appendix B includes examples of solar energy goals, objectives, and policies in comprehensive plans from communities across the country.

▶ PINAL COUNTY, ARIZONA

Pinal County, Arizona, is located between Phoenix and Tucson. It is the third most populous county in Arizona with a 2010 population of 375,770, and it was the second fastest growing county in the U.S. between 2000 and 2010. The county's most recent comprehensive plan includes references to solar energy use in its existing conditions assessment and its goals, objectives, and policies. Furthermore, the county considered solar energy use when amending its future land use map.

According to Steve Abraham, AICP, Planning Manager with Pinal County, a confluence of events led to Pinal County's decision to incorporate solar energy use into its latest comprehensive plan. First, the state legislature passed a law requiring that the comprehensive plan include an energy element and that the element identify policies that encourage and provide incentives for the efficient use of energy while assessing policies and practices that provide for greater uses of renewable energy sources (§11-804). Additionally, in 2006 the Arizona Corporation Commission approved the Renewable Energy Standard and Tariff, which requires regulated utilities to generate 15 percent of their energy from renewable sources by 2025 (ACC 2013). Then, in 2008, the county began the process of updating its comprehensive plan. Throughout the year-and-a-half-long public input process, a strong contingent of residents consistently advocated for sustainability principles. In response, the county's board of supervisors created the Sustainable Pinal program and appointed the Sustainable Pinal Citizen Task Force. The board charged this task force with making recommendations on matters including energy conservation. Finally, throughout this period, private companies were regularly approaching the county with questions about installing solar energy systems (Abraham 2013).

The county's comprehensive plan discusses solar energy use in two different chapters. First, Chapter 3 of the plan addresses "Planning Guidelines" that provide direction and guidance in developing or reviewing commerce-

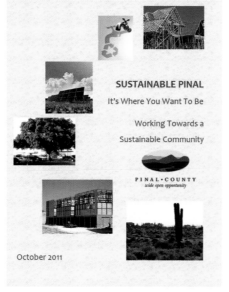

The Sustainable Pinal Citizen Task Force delivered its recommendations for creating a successful sustainability program in Sustainable Pinal: It's Where You Want to Be.

Source: Pinal 2011

related projects. This chapter's agriculture planning guidelines state that solar and wind energy generation and other renewable energy production should be viewed as compatible with the county's farming heritage. The bulk of the solar energy discussion can be found in Chapter 7, the environmental stewardship chapter of the plan. The chapter is broken into three strategic areas: conservation, renewable energy sources, and energy generation and transmission. The conservation section stresses the importance of energy conservation and its various benefits and implications. The renewable energy sources section describes how the county will provide support for making renewable energy more feasible and attractive through regulatory and taxation policies, by ensuring enough space is available for the siting of future facilities, and through supporting education and training opportunities. The energy generation and transmission section discusses the existing energy providers and future electrical energy needs of the county.

Table 4.1 shows the goals, objectives, and policies relating to solar energy included in this chapter.

Table 4.1. Solar energy goals, objectives, and policies in the Pinal County, Arizona, comprehensive plan

Source: Pinal 2009

Goal 7.3: Improve the energy efficiency of Pinal County government.

Objective 7.3.1: Set an example by improving energy efficiency and use of renewable sources in County facilities, vehicle fleets, and equipment.

Policy 7.3.1.3: Locate solar energy generation equipment on County facilities which cost/benefit analyses prove advantageous.

Goal 7.4: Improve the energy efficiency of structures in Pinal County.

Objective 7.4.1: Improve the energy efficiency of new construction and the existing building stock through building codes and processes.

Policy 7.4.1.2: Encourage the expansion of energy efficient building practices.

Policy 7.4.1.4: Support refurbishing and remodeling projects to include energy efficiency components through expedited permitting and assistance.

Objective 7.4.2: Reduce energy demand through community design.

Policy 7.4.2.1: Encourage developments that use energy smart site design (e.g., solar orientation, cluster development).

Goal 7.6: Expand renewable energy in Pinal County.

Objective 7.6.1: Support small scale renewable energy projects.

Policy 7.6.1.3: Work with developers and energy providers to design neighborhoods with optimum solar orientation.

Policy 7.6.1.5: Develop/amend ordinances to protect solar access through sensitive building orientation and for property owners, builders and developers wishing to install solar energy systems.

Additionally, the county designated several "Employment" and some "General Public Facilities/Services" areas on the Future Land Use Map, for which the county is amenable to a rezoning to accommodate large-scale solar installations if approached by a utility or developer. Since the adoption of the comprehensive plan, the county has entitled two large-scale solar energy facilities in remote, undeveloped areas of the county.

Currently, the county's development regulations only specifically address accessory solar energy systems, including definitions, general requirements, and uses permitted. Accessory solar energy systems are allowed by right in residential zones subject to certain requirements (§2.210). There are no specific definitions or requirements for large-scale solar facilities. They are considered power plants and are regulated as such. Power plants are permitted by right in the Industrial Zoning District (I-3), and applicants must obtain a special use permit in all other zoning districts.

According to Abraham, the Sustainable Pinal Citizen Task Force has recently proposed an amendment to allow plan updates for primary-use solar installations (regardless of size) to be approved concurrently with rezoning. This amendment would be significant because major amendments are considered by the county board on an annual basis. According to Abraham, there has been no major resistance to this recommendation at either the supervisor or community level. He believes that if this amendment passes, it will likely stimulate a series of additional amendments related to solar energy systems in the county's development regulations.

SOLAR IN SUBAREA PLANS

Subarea plans are plans that include goals and objectives for a discrete geographic area within a jurisdiction. Some common types of subarea plans include plans for specific sectors, neighborhoods, corridors, or special districts, such as transit station areas, redevelopment areas, or areas designated for historic preservation. These plans may cover a wide range of topics relevant to the plan area, essentially functioning as smaller-scale comprehensive plans, or they may be strategic in nature, focusing on a subset of topics with special urgency.

The limited extent of subarea plans has both advantages and disadvantages. Because comprehensive plans can seem abstract or diffuse to residents, business owners, or institutions that identify more with specific neighborhoods than with a city as a whole, planners often have an easier time identifying and engaging key stakeholders when a plan has clear implications for these stakeholders' homes, businesses, and shared public spaces. The other clear potential advantage of subarea plans is that these plans can be more specific about how goals and objectives apply to individual parcels of land. On the flip side, strong emotions can lead to a loss of objectivity, making it difficult for communities to prioritize scarce resources.

When considering the limited extent and greater specificity of subarea plans in the context of planning for solar energy use, plan authors have opportunities to discuss the neighborhood- or parcel-level implications of policies and actions aimed at increasing adoption of solar technologies. Subarea plans can provide greater detail about preferred locations for solar installations and go into more depth about the regulations, incentives, and potential competing interests that may either support or inhibit local solar market growth.

Many communities incorporate design guidelines for future development into subarea plans. For example, both Austin, Texas, and Amherst, Massachusetts, have adopted neighborhood plans that address solar design. Austin's Brentwood/Highland Combined Neighborhood Plan recommends subdivision layouts and lot configurations that maximize solar access, and it encourages concentrating windows on the south face of buildings to promote passive solar heating. Similarly, Amherst's Atkins Corner plan includes a workbook of sustainable development design options, which highlights the importance of passive solar design as part of an overall strategy to maximize climate-friendly development.

▶ MONTGOMERY COUNTY, MARYLAND

Montgomery County, Maryland, situated just north of Washington, D.C., and southwest of Baltimore, is a national leader in land-use planning and smart growth. Over the past several decades, the county's planning department has developed numerous subarea plans to implement its "wedges and corridors" growth management strategy. One of the department's latest innovations is a new sustainable neighborhood planning tool: a set of principles for reviewing plans and project proposals with the express intent of reducing energy consumption and environmental

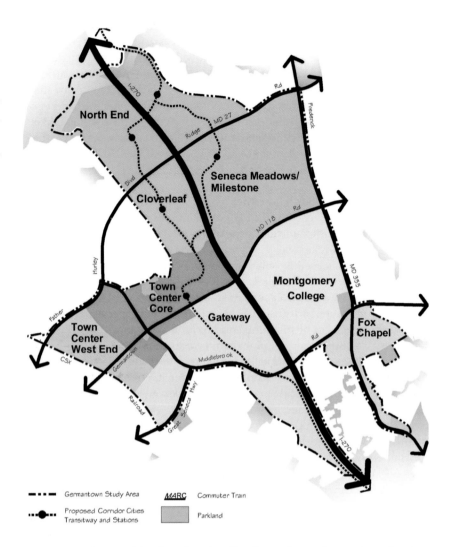

Figure 4.2. *The Germantown Employment Area Sector Plan envisions seven transit-oriented districts, each with a mixed-use core.*

Source: MNCPPC 2009

impacts at the neighborhood scale. The first plan to be reviewed using these principles is the Germantown Employment Area Sector Plan, a subarea plan for 2,400 acres in the county's central employment corridor (MCPD 2013) (Figure 4.2).

The Germantown Employment Area plan establishes a vision for transforming an auto-oriented corridor into a vibrant town center surrounded by mixed use districts. Key recommendations include on-site renewable energy production and green neighborhood design and building techniques to conserve energy, conservation of bulk and mass in building design to improve light on the street and minimize shadows, and development in an urban pattern that allows for creative design and building options that enhance environmental quality (MNCPPC 2009). Design guidelines developed as a part of the plan include designing for solar orientation of streets, public spaces, and buildings; using green or cool roof technologies; and incorporating measures such as day lighting, LED street lights, and solar panels that promote energy efficiency (MNCPPC 2010).

The planning department's new sustainable neighborhood planning tool originated with a staff desire to take a more active role in promoting sustainability at the neighborhood scale. Staff researched the limited examples of communities that were doing sustainable neighborhood planning and combined principles and precedents from these initiatives with components of the LEED for Neighborhood Development rating system and others for sustainable neighborhood planning to develop a tool that could be used by the planning board during project review. Staff presented its research and the associated tool at a March 2013 briefing with the board, which unanimously accepted the tool for use during plan review (Carter 2013).

The tool's neighborhood-scale review principles fall under four broad categories:

- Neighborhood location and place making

- Linkages and pedestrian orientation

- Energy conservation, solar orientation, and green buildings

- Environmental protection and conservation

The energy conservation, solar orientation, and green buildings category includes consideration of block and street orientation, building height and shading, site planning for solar orientation, and building orientation for day lighting of neighborhoods (MCPD 2013). Collectively, the principles create a new framework for the planning and design of environmentally superior neighborhoods. So far the planning department's sustainable neighborhood planning approach has been applied to two project plans in the Germantown Employment Area and a large plan in the Cabin Branch community of Clarksburg (Carter 2013).

SOLAR IN FUNCTIONAL PLANS

Functional plans are standalone plans for systems or special topics that have spatial planning implications but are not, fundamentally, rooted in a single subarea of a community. Examples include capital improvement plans, affordable housing plans, transportation system plans, and open space network plans. As an increasing number of communities acknowledge the importance of energy and climate planning, other functional plans—such as sustainability plans, climate action plans, and energy plans—have become increasingly common. Some communities use functional planning processes as ways to incrementally create or update comprehensive plans. Other communities create functional plans either to address new topics rising on the public agenda or in response to special federal or state funding requirements.

While there are numerous local sustainability, climate action, and energy plans that incorporate goals, policies, and actions related to the promotion of solar energy, these plans seldom hold the same statutory authority as the comprehensive plan. Therefore, communities should incorporate relevant policies and action items related to solar energy use from these functional plans into the comprehensive plan. This may involve incorporating functional plans into the comprehensive plan by reference, or it may mean updating specific sections of the comprehensive plan to reflect new community priorities. Appendix C of this report includes sample solar strategies, measures, and action in functional plans.

As an example, in 2010 Lawrence Township, New Jersey, adopted a Green Buildings and Environmental Sustainability plan as an update to the township's 1995 Master Plan. The new plan element contains an explicit reference to New Jersey's planning enabling law, discusses existing conditions relating to solar energy production, and lays out several specific goals, objectives, and policies for promoting passive solar design and solar energy systems in both private development projects and municipal facilities.

▶ ANN ARBOR, MICHIGAN

Ann Arbor is located in southeast Michigan, about 40 miles west of downtown Detroit. It is the sixth largest city in Michigan and is home to the University of Michigan. In 2010 the city approved a functional plan devoted exclusively to strategies for promoting solar energy use (Ann Arbor 2010). Although development of this solar plan was made possible because of funding received through the U.S. Department of Energy's (DOE) Solar America City initiative, Ann Arbor used the opportunity to build upon previous energy initiatives, and since its adoption, the city has worked continuously to implement the plan's recommendations.

XSeed Energy is a grassroots initiative in which residents, businesses, and others work together to develop highly visible renewable energy projects. In September 2013 XSeed Energy placed its first installation (3 kW) on the landmark Michigan Theater in downtown Ann Arbor, Michigan.

Nathan Geisler

According to Nate Geisler, Energy Programs Associate with the city's Energy Office, Ann Arbor's interest in energy conservation and renewable energy sources can be traced back to the energy crisis of the 1970s. The city developed an Energy Plan in 1981, created an Energy Office in 1985, and in the intervening years also established an Energy Commission (Geisler 2010). The Energy Plan established goals and programs to reduce energy use and costs while moving the city towards more sustainable energy use (Ann Arbor 2013a). The city council approved the plan in 1981 and an update in 1994. The Energy Office is a part of the Systems Planning Unit that is responsible for long-range asset management planning for the city (Ann Arbor 2013b). The Energy Commission, which is appointed by the mayor and meets monthly, is charged with several tasks including helping oversee city policies where energy efficiency and renewable energy should be addressed, advising the city council, and creating reports and recommendations about municipal and community energy efficiency and renewable energy (Ann Arbor 2013c).

Based on a recommendation in the Energy Plan, the city established a Municipal Energy Fund in 1998 to fund energy-efficient retrofits at city facilities in an attempt to continually reduce operating costs over time. The city invests these funds in a variety of energy projects, including solar energy projects (Ann Arbor 2013d). Additionally, in 2005, Mayor John Hieftje issued a Green Energy Challenge. The current challenge calls for 30 percent of energy for municipal operations—and 5 percent for the whole city—to be green by 2015 (Ann Arbor 2013e).

In 2007 the DOE designated Ann Arbor a Solar America City. As a result, the city received funding to integrate solar energy throughout the community. In October 2010, with assistance from the Clean Energy Coalition, the city produced a Solar Plan built on the city's previous initiatives, including the Energy Plan and Green Energy Challenge. The plan addresses the benefits of and barriers to solar, provides an energy profile for the city, and identifies best practice strategies to increase the adoption of solar energy technologies. It culminates in eight recommendations, and includes information on why each recommendation was included and how it can be achieved (Ann Arbor 2010).

The city has begun to successfully implement the recommendations set forth in the Solar Plan. For example, to achieve Recommendation 1, Commit to a Solar Plan Implementation Process, the city first identified the need to prioritize recommendations. It obtained a $95,000 grant from The Home Depot Foundation to undertake a sustainability framework project designed to capture, organize, and prioritize all of the goals, objectives, policies, and ideas listed in each of the city's more than 25 plans, including those found in the Solar Plan (Geisler 2013). The project also includes development of a sustainability action plan to connect the overarching goals with quantifiable targets. Through this process, the city realized that its plan goals and recommendations could be divided into four primary areas: climate and energy, community, land use and access, and resource management. In February 2013, city council approved a resolution incorporating 16 sustainability goals in the City Master Plan, including three specific to climate and energy (Ann Arbor 2013f).

The city has also worked to achieve Recommendation 2, Design Municipal Solar Financial Incentives, by creating Michigan's first Property Assessed Clean Energy (PACE) Program. According to Geisler, the city used its Energy Efficiency and Conservation Block Grant to establish the PACE District and program policies. PACE is a voluntary special assessment which can be levied on commercial parcels for the purpose of financing energy efficiency or renewable energy projects.

The city also offered a set of small commercial energy revolving loans to implement efficiency upgrades of small commercial buildings, including a solar project at a downtown brewery (Ann Arbor 2013g). Additionally, the city has a grassroots initiative, the XSeed Energy program (www.xseedenergy.org), which is a community donation–driven solar program started as part of the Solar America Cities project that is helping to get solar on downtown buildings like the Michigan Theater, a major destination and landmark in the community (Clean Energy Coalition 2013).

SUMMARY

Plan making is the second strategic point of intervention for communities looking to promote solar energy use through planning. Most communities adopt local plans to articulate how specific regulations, programs, investments, and actions should be used to help actualize their long-term visions for the future. Planners and public officials then use these plans to inform decisions that affect the social, economic, and physical growth and change of their communities. Plan authors have opportunities to discuss or encourage solar energy use throughout local comprehensive, subarea, and functional plans. And plans that include background information about local solar markets along with solar-supportive goals, objectives, policies, and actions send clear signals to residents, business owners, and other community stakeholders about where and how solar energy use will be sanctioned or encouraged locally.

CHAPTER 5

Regulations and Incentives

Ann Dillemuth, AICP, Darcie White, AICP, Paul Anthony, AICP,
Justin Barnes, David Morley, AICP, and Erin Musiol, AICP

 The third strategic point of intervention for communities looking to promote local solar energy use through planning is regulations and incentives. One of the keys to solar development is a supportive regulatory environment. At the local level, development regulations may support or pose barriers to solar energy use. In many instances, planners are well positioned to examine local zoning, subdivision, and building codes to determine if—and how—they address solar energy, and whether barriers, either intentional or unintentional, exist. Beyond simply enabling solar energy use, communities can also use development and financial incentives to actively promote solar development.

This chapter begins with a discussion of the different types of zoning, subdivision, and building code provisions that communities can use to enable different types of solar development. The subsequent sections provide an overview of the different potential developments and financial incentives localities may employ to encourage private investment in solar energy systems.

DEVELOPMENT REGULATIONS

While the primary role of local development regulations (i.e., zoning, subdivision, and building codes) is not to create an incentive for a particular type of development, clear definitions, use permissions, and development standards for solar development can have this effect. This is because explicit regulatory language sends a signal to developers and installers that the community is prepared to accommodate solar development—that it is "open for solar business." Predictable development review and approval processes can also help developers secure project financing.

In terms of creating a supportive regulatory environment, local development regulations serve a number of important functions: clarifying what types of solar development are allowed and where; mitigating potential nuisances associated with solar equipment, such as visual impacts or encroachment; and addressing solar access issues. Beyond these basic functions, communities may also use development regulations to encourage or require developers to orient new streets and lots in ways that maximize solar access and builders to construct new solar-ready homes.

As of February 2013, 39 states and the District of Columbia have adopted one or more types of solar access laws (DSIRE 2013c). These include laws that (1) preempt local development regulations or homeowners' association conditions, covenants, and restrictions that prohibit solar energy systems; (2) enable solar easements, which allow a landowner to enter into an agreement with an adjacent landowner to ensure that sunlight reaches the property; (3), authorize local zoning authorities to adopt solar access regulations, which permit local authorities to adopt zoning provisions that preserve solar access; and (4) enable solar shade-protection regulations, which ensure that the performance of a solar energy device will not be compromised by shade from vegetation on adjoining properties (Kettles 2008). Thus, planners should always check state law to know precisely what authority has been given (or not given) to local governments to regulate solar energy systems.

While specific regulations will vary from community to community based on local goals and context, there are a number of common types of provisions related to different aspects of solar energy use. The second section of Appendix A in this report contains a series of questions for community stakeholders to consider when auditing existing or proposed development regulations. Appendix D includes a model solar development regulation framework, while Appendix E provides numerous examples of model solar development ordinances.

Assessing Potential Barriers to Passive Solar Design

Common passive solar design techniques are typically permissible under local building, subdivision, and zoning codes. However, some development standards can pose barriers to passive solar design. And assessing the effects of existing standards on passive solar design options is an important step for communities interested in creating a more solar-friendly regulatory environment.

Generally speaking, prescriptive standards related to landscaping and site or building design have the greatest potential to frustrate efforts to implement specific passive solar techniques. For example, some communities stipulate where required landscaping must be provided on a particular parcel. If these prescriptive standards are based purely on aesthetics, they may have the unintended consequence of limiting the efficacy of certain day lighting or space heating techniques. Similarly, prescriptive standards that stipulate specific setbacks, bulk and massing, and architectural features for buildings may, unintentionally, prevent building placements that would optimize day lighting or space heating.

Given that most local development regulations are proscriptive (i.e., creating a permissible buildable envelope) or performance-based (i.e., setting an environmental performance target), communities with prescriptive standards have, typically, arrived at those standards through a deliberative process. As a consequence, communities that have identified potential barriers to passive solar design due to prescriptive standards will need to balance a desire to promote a specific built form or community character against a desire to promote solar energy use. Possible compromises include a limited exception from certain development standards or a ministerial (rather than a quasi-judicial) variance procedure for projects incorporating passive solar design techniques.

Establishing Clear Definitions and Use Permissions

Many zoning codes fail to define specific terms related to solar energy use and do not clearly identify the zoning districts in which solar energy systems are allowed. This silence often creates uncertainty about the legality of various types of solar development, forcing local officials to make ad hoc determinations about the similarity of specific installations with other defined uses. When a community adds definitions and use permissions to its code, it eliminates this inadvertent barrier to solar development.

Some communities choose to distinguish between solar thermal and solar PV systems, but many others use "solar energy system," or a similar broad term, to refer to any type of solar collector and its associated equipment. In the same way, some communities define and classify large-scale or primary-use solar installations as a distinct use, while others simply distinguish between accessory and primary-use installations in tables or lists of uses permitted by district.

Many communities permit accessory solar energy systems (see Chapter 2) by right in all zoning districts. When localities explicitly acknowledge primary-use systems (see Chapter 2) in their codes, they often permit these installations either by right or through a discretionary approval process (e.g., a conditional use permit) in rural or industrial districts. Some communities, though, have taken a more permissive approach by permitting primary-use systems in a wide range of residential and nonresidential districts. For example, Milwaukee's zoning code permits primary-use installations (defined as "solar farms") by right in all residential and many nonresidential districts and through a discretionary review in most other districts (see Subchapters 5–10 of Chapter 295 of the city's Code of Ordinances).

Even in cases where zoning codes explicitly address solar energy systems, subtle barriers such as height restrictions, lot coverage limitations, and setback, screening, landscaping, and utility requirements may still impede solar development. In response, many communities provide limited exceptions to certain dimensional or development standards for solar systems. For example, Hermosa Beach, California, allows solar energy systems to exceed height limits to the minimum extent necessary for safe and efficient operation and provides flexibility in modifying other development standards that might reduce system performance (§17.46.220).

Beyond definitions, use permissions, and limited exceptions to general development standards, a number of communities have added use-specific standards for different types of solar development. These additional standards can mitigate concerns related to neighborhood character and help avoid conflicts over competing values, such as tree protection or historic

Land Use	Zone District																Supplemental Standards
	Residential								Mixed Use					Commercial and Light Industrial			
	R-1-43	R-1-18	R-1-12	R-1-9	R-1-6	R-2	R-MF	R-MH	M-N	M-G	M-C	M-E	M-R	C-R	L-I	LI-RD	
P = Permitted A = Accessory S = Special L = Limited [blank] = Prohibited																	
Public/Civic/Institutional																	
Solar Garden														P	P	P	
Other (continued)																	
Solar Collection System	A	A	A	A	A	A	A	A	A	A	A	A	A	A	A	A	See Section 17.5.5.3

In Lakewood, Colorado, accessory solar PV and thermal systems are permitted by right in all zoning districts, subject to use-specific standards. The city also permits solar gardens (primary-use solar energy systems with a rated capacity less than 2 MW) by right in commercial and industrial districts.

Source: Lakewood 2012

preservation (see Chapter 3). Consequently, use-specific standards give communities the confidence to permit solar energy systems by right in a wider variety of zoning districts. This is important because the uncertainty associated with discretionary approvals increases costs for owners and developers and may make project financing difficult. Finally, use-specific standards can help set the stage for future installations by ensuring that new development is situated for maximum solar access and new structures are wired and plumbed for solar electric and hot water systems.

Accessory Use Standards for Solar Energy Systems

Accessory use solar energy systems typically have minimal impacts. Rooftops provide a vast amount of potential space for solar development that does not consume new land or increase impervious surface area within a jurisdiction. The main concerns related to accessory solar energy systems are aesthetic: how and where systems are placed on a property.

Consequently, use-specific standards often address placement. Some codes encourage rooftop over freestanding systems. Many require rooftop installations to be located on side or rear roof slopes rather than street-facing roof slopes, when possible, for aesthetic reasons. Similarly, some codes limit the height that rooftop collectors may extend above the roofline (often two or three feet); alternatively, they may exempt solar collectors altogether from district height restrictions, along with other typical rooftop structures such as chimneys, air conditioning units, or steeples.

For freestanding systems, communities often restrict placement to side or rear yards and sometimes require screening from public rights-of-way. Many codes also address system appearance, requiring neutral paint colors and screening of nonpanel system components. In all placement and screening considerations, however, the effects of requirements on system function

A grid-connected solar energy system on a residential rooftop in Gardner, Massachusetts

Bill Eager (NREL 00568)

must be considered, and most codes provide for some degree of flexibility to ensure that property owners can work within site and structural constraints to achieve reasonable solar collection.

While accessory solar energy systems are typically installed to meet on-site power needs for buildings and other uses, there is no need to place limitations on the size or power production capacity of an accessory system—height and location restrictions will place reasonable constraints on the size or extent of panels and their placement. In most states, net-metering arrangements allow solar energy system owners to feed excess energy back into the grid. Adding stipulations that accessory systems be limited in capacity to on-site power needs or implementing an arbitrary system size cap can add unnecessary barriers to solar implementation.

Finally, some communities require that solar systems remain well-maintained throughout their working life, and they require the decommissioning of collectors once they cease to function properly or if they are abandoned for a certain length of time. This ensures safety and prevents obsolete or damaged systems from becoming public eyesores or nuisances. Appendix F contains multiple examples of communities with use-specific standards for accessory solar energy systems.

Primary Use Standards for Solar Energy Systems

For primary-use solar energy systems, system size (whether measured by capacity or by physical space requirements) is an important factor as communities consider the appropriate scope and level of detail for use-specific standards. These installations can range in size from less than an acre in urban settings to hundreds or even thousands of acres in remote locations.

Large solar farms can raise concerns regarding impervious surface coverage, tree and habitat loss, transmission infrastructure, and construction impacts. Furthermore, proposals for large installations are more likely to court controversy, especially when developers propose greenfields or productive agricultural lands as sites.

Common use-specific standards for solar farms include height limitations, setbacks from property lines or neighboring structures, and screening from adjacent public rights-of-way. For security and safety reasons, many communities require that solar farms be securely fenced, that warning signs be posted, and that on-site electrical interconnections and power lines be installed underground.

Many communities require a site plan review for a large installation as well as an agreement with a utility for interconnection of the completed facility. Some localities also require stormwater management plans and, in more rural communities or areas that abut public land, environmental analyses for

Agricenter International's solar farm in Shelby County, Tennessee, provides enough electricity to power hundreds of homes.

Thomas R. Machnitzki / Creative Commons 3.0

potential impacts on wildlife and vegetation. Finally, communities often require owners to decommission nonfunctioning facilities, and some localities require restoration of the site to its previous condition, especially for formerly agricultural lands. An example of a community with extensive standards for solar farms is Iron County, Utah, which has adopted code provisions addressing siting considerations and requirements for analyses of local economic benefits, visual impacts, and environmental impacts, in addition to transportation plans for construction and operation phases (Chap. 17-33).

In contrast, smaller solar gardens may have impacts more closely analogous to freestanding accessory-use systems. As stated in Chapter 2, many smaller primary-use installations are community solar projects. In addition to being less conspicuous than solar farms, community solar gardens tend to be less controversial because the benefits of these projects are often clear to nearby residents and business owners.

Although relatively few communities have adopted use-specific standards for solar gardens, those that have tend to focus more on neighborhood compatibility rather than communitywide impacts. For example, Boulder County, Colorado, permits solar gardens with capacities of less than 100 kW in all districts by ministerial site plan review and those with capacities up to 500 kW in all districts by discretionary review, subject to specific provisions addressing visual impact and glare on adjacent properties (§4-514.L).

In more urban settings, drawing distinctions between solar farms and solar gardens may be less important. This is because district development standards and the prevailing land economics of the community will act in concert to limit the overall size of any single project.

▶ GRANVILLE COUNTY, NORTH CAROLINA

North Carolina's Granville County is located in the northern piedmont region on the Virginia border. As a result of the events described below, in 2007 staff for the largely rural county's planning department began noticing an increase in the number of inquiries about installing renewable energy systems (Baker 2013). In response to this interest, the county has updated its development regulations to acknowledge and sanction different scales of solar development in various zoning districts throughout the jurisdiction.

With the signing of Session Law 2007-397 in August 2007, North Carolina became the first state in the southeast to adopt a Renewable Energy and Energy Efficiency Portfolio Standard (REPS). This law requires investor-owned utilities to meet up to 12.5 percent of their energy needs through renewable energy sources or energy-efficiency measures (NC Utilities Commission 2013). Meanwhile, gas prices had skyrocketed and the planning department had been receiving inquiries from residents, especially those in the agricultural community, about promoting the use of biodiesel and ethanol to operate vehicles and tractors (Baker 2013).

These events made staff realize that they would likely experience a growth in the number of requests for renewable energy systems and facilities in the county. They began to think about how to address renewable energy systems and facilities in their code of ordinances.

Luckily, the state biofuels research center, the Biofuels Center of North Carolina, is located in the county seat of Oxford. The planning department reached out to the center to obtain information and resources on biodiesel and ethanol. They also searched the North Carolina Solar Center resources for examples of local codes and information about available incentives and conducted a national search for sample and model ordinances to use as inspiration (Baker 2013). During this search, the county found the solar development ordinance in From Policy to Reality: Model Ordinances for

Sustainable Development, a set of model ordinances aimed at communities in Minnesota, particularly useful (MEQB 2013).

In 2009, after thorough research, Granville planning staff undertook a two-step process to amend its code of ordinances to include alternative energy sources. The planning board and county commissioners were both very receptive to the amendments. The first amendments related to renewable energy were adopted in August 2009. Additional amendments as well as revisions to earlier amendments were incorporated in August 2010 (Baker 2013).

The amended code now includes definitions for various terms related to solar energy use, such as solar energy, photovoltaic system, solar energy system, solar farm, and solar mounting devices (§32-1331). It permits accessory solar energy systems by right in all zoning districts, subject to specific setback and height requirements (§32-162, §32-163). Additionally, it includes height, setback, and visibility requirements for freestanding solar energy systems and requires approved solar components and compliance with building and energy codes (§32-233). Finally, the code permits primary-use solar farms by right subject to certain requirements (location, design, height) in the General Industrial District (I-2), with conditional use permit approval by the board of adjustment (public hearing is required) in the Prime Industrial District (I-1), and with special use approval by the planning board and board of commissioners (public hearing is required) in the Agricultural Residential District (AR-40) (§32-142).

The initial amendment did not permit solar farms in agricultural districts. After the county was approached by a solar developer, it learned of a need for land for solar farms. Staff initiated an amendment to the newly adopted revisions to allow solar farms with special use approval in the AR-40 District. They paid particular attention to screening to ensure that solar farms will blend into the rural character of the district. Screening requirements state that solar farms must be set back at least 25 feet and are subject to buffer standards that screen them from routine view from public rights-of-ways and adjacent residentially-zoned property (§32-264). As a result of this amendment, the county approved its first solar farm in early 2012. The solar farm utilizes 27 acres of a 40-acre agriculture site. Hay production still occurs on the remaining portion of the site. The 1.9 MW farm, built by Sun Edison, has been operating since October 2012 and sells its electricity to local utility Progress Energy (Baker 2013).

Solar Access Protections

As noted in Chapter 3, there is currently no common law "right" to sunlight for solar energy production in the U.S.; solar access is protected only in states that have passed solar-rights statutes or in jurisdictions where local governments have created solar protections via ordinance. Local solar access provisions guarantee property owners a reasonable amount of sunlight and protect installed systems from being shaded by structures and vegetation. Communities can provide for the protection of solar resources in three main ways: (1) solar easements, (2) solar access permits, and (3) solar "fences."

A solar easement protects a property owner's access to sunlight through negotiated agreements with neighboring owners that are recorded with the appropriate authority. As noted above, some states have enacted statutes defining and enabling local solar easements; one example is New Hampshire, which offers a model "Solar Skyspace Easement" template in its state statutes (§477:51). Apart from state enabling laws, some local governments authorize or require solar easements for new solar energy systems in order to minimize the potential for future conflicts. Because this mechanism puts the onus for securing an easement on the property owner hosting the system, it is typically the least contentious local approach to protecting solar access. For the

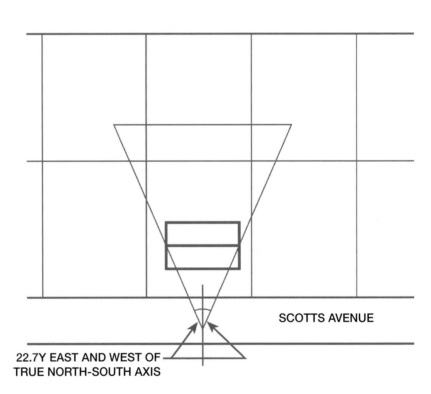

These figures from Clackamas County, Oregon's Zoning and Development Ordinance illustrate how solar access provisions can provide detailed mechanisms for establishing and protecting access to sunlight for individual lots.

Source: Clackamas 2013

same reason, solar advocates and other stakeholders often perceive solar easements to be the weakest form of protection.

A solar access permit protects access to the solar resource using a slightly different mechanism: a property owner provides documentation of a solar energy system to the local government and obtains a permit providing protection from shading caused by future construction or tree growth on neighboring properties. To balance the rights of other property owners, communities may allow for some degree of system shading above a threshold that ensures the system's effectiveness will not drop below a certain percentage. For example, the Village of Prairie du Sac, Wisconsin, allows owners to obtain a solar access permit to protect their solar energy systems from "impermissible interference," including shading of more than 95 percent of collector surface between 9:00 a.m. and 3:00 p.m. each day (§10-8). Solar access permits bring solar access protection into the realm of code enforcement, taking some of the burden off of system owners or hosts. For this reason, solar advocates and other stakeholders often perceive these permits as a stronger form of protection than solar easements.

Solar fences protect access universally, whether or not solar development exists or is planned for a specific lot. Consequently, solar advocates and other stakeholders often perceive this approach to be the strongest form of solar access protection. Communities establish solar fences for designated lots in the initial subdivision process, by delineating an imaginary box on each lot within which sunlight must fall unobstructed by neighboring structures or vegetation, often for a certain daily amount of time (commonly defined as, at minimum, between two to three hours on either side of noon on the winter solstice). Neighboring property owners are prohibited from erecting any structures that would cast shadows during that time in the lot area protected by the ordinance. For example, Fort Collins, Colorado, limits the shading of structures on adjacent property to that generated by a 25-foot hypothetical wall located along the property line, but exempts certain high-density zoning districts from this provision (§3.2.3).

Boulder, Colorado, uses this approach to establish three different solar access (SA) areas that balance solar access with restrictions on development density and height (§9-9-17). The code provides for solar fences in the first two SA areas that set maximum allowable shading of lot building envelopes; in the third SA area, solar protections are only granted for specific properties through permits. Solar access area designations may be amended by property owners through a public hearing and review process. See below for more information on Boulder's solar access regulations.

▶ BOULDER, COLORADO

Boulder, Colorado, located about 30 miles northeast of Denver at the base of the Rocky Mountains, is the 11th most populous city in the state and home to the University of Colorado at Boulder. Boulder was one of the first communities in the U.S. to adopt a solar-fence access protection ordinance.

In response to the diminishing supply and increasing cost of conventional energy resources, Boulder adopted a solar access ordinance to protect the use of solar energy in 1982. This ordinance is incorporated into the city code (§9-9-17) and has remained virtually unchanged since its adoption.

Boulder's solar access ordinance guarantees access to sunlight for home-owners and renters in the city by establishing a solar fence for each lot. Boulder is divided into three solar access areas: Solar Access Area I (SAA I), Solar Access Area II (SAA II), and Solar Access Area III (SAA III). Each solar access area affords protection for a defined period of time each day. The degree of protection afforded through the ordinance is based on the property's solar access area designation.

SAA I regulations apply in low-density residential zoning districts composed of primarily detached single-family homes. The regulations are designed to protect solar access principally for south yards, south walls, and rooftops through establishment of a 12-foot hypothetical solar fence on the property lines of the protected buildings. SAA II applies to higher-density residential and mixed use zoning districts, and is designed to protect solar access principally for rooftops through the establishment of a 25-foot solar fence. All remaining zoning districts, including the downtown and most commercial and industrial districts, fall into SAA III (Holmes 2013). In this solar access area, solar access protection is available to property owners through a solar permit which limits the amount of permitted shading by new construction.

When applying for a building permit, applicants must submit an adjusted shadow analysis for review to the planning and development services department as part of the development review process. The shadow analysis includes a solar access site plan that determines the height of the shadow-casting portion of the roof and the approximate shadow cast by the proposed structure (Boulder 2006). If the shadow cast is entirely within the property lines, the structure is in compliance. If the shadow analysis indicates that the shadow is close to or within one foot of the shadow of the solar fence, the department may require height verification as a condition of building permit approval (Holmes 2013). Height verification calculations must be performed by a licensed surveyor (Boulder 2012). If the shadow cast falls outside of the property lines, the applicant must either redesign the project or apply for a solar exception. A solar exception can be granted administratively if there are no objections from affected property owners and the application complies with criteria outlined in the code (§9-9-17.f). If an affected property owner objects, or if staff finds the proposal does not meet the criteria for a solar exception, a public hearing before the Board of Zoning Adjustment is required.

Although Boulder is largely built out, the ordinance also includes requirements for solar siting in new construction. It requires the roof surfaces of all units in new planned unit developments and subdivisions (both residential and nonresidential) to be oriented within 30 degrees of a true east-west direction, to be flat or not sloped towards true north, to be able to physically and structurally support at least 75 square feet of unshaded collectors per dwelling unit, and to have unimpeded solar access (§9-9-17.g).

The ordinance also addresses conflicts due to vegetation. The ordinance does not apply to existing vegetation, but if vegetation is planted after the time an application is filed and it shades or grows to shade solar access, the vegetation owner may be asked to remedy the shading at their own cost (§9-9-17.h.14).

Property owners in SAA III, or those property owners who have installed or plan to install a solar energy system in SAA I or SAA II and need more protection than is automatically provided, can apply for a solar access permit (§9-9-17.h). According to Brian Holmes, Zoning Administrator with the City of Boulder, this is the least-used part of the ordinance. The city has not received an application for a solar access permit since the 1980s. Although infrequently utilized, the ordinance specifies eligibility standards and application requirements for interested applicants.

Holmes (2013) believes the solar access ordinance is successful because property owners feel confident that their investments in solar energy systems will be protected. They know that the city has a process in place and that solar access is considered during every development review. As a result, there have been few concerns or complaints from residents.

Although the ordinance itself has undergone few amendments, the department is continuously working to improve the shadow analysis submission process. Staff members undergo training to ensure they can answer applicant questions at the counter, and the city accepts analyses ranging from sketches done by homeowners to professional drawings by architects and engineers (Holmes 2013).

Solar Site Design Requirements

Most existing development patterns and site layouts do not protect or take advantage of solar resources. A number of communities have added solar site design provisions to their subdivision codes or general site development standards to ensure that future development is optimally sited for solar use. These provisions sometimes go hand-in-hand with solar access requirements.

Solar site design provisions set standards for lot size and orientation—as well as site layout for parcels—that provide for the construction of buildings whose southern sides or ends have unobstructed solar access for a designated time during each day (as in the case of solar access ordinances, typically a minimum of two to three hours on either side of noon on the winter solstice). Requirements include street and lot orientation within certain degrees of an east-west axis to ensure adequate sunlight access. Typically, a certain minimum percentage of lots within new subdivisions must comply with these requirements. Solar site design provisions may also place restrictions on the height and location of structures on the lot so that basic solar access to neighboring lots will not be blocked, or they may allow flexibility within setback regulations to maximize solar access for new construction. Such provisions do not only benefit homeowners who choose to purchase and install solar energy systems, but also maximize opportunities for the design of passive heating and cooling features.

To complement its solar access provisions, Boulder requires new residential development to have roof and exterior wall surfaces that are oriented toward the sun, have unimpeded solar access, and are structurally capable of supporting solar collectors. Similarly, Laramie, Wyoming, requires at least 40 percent of lots less than 15,000 square feet in area in single- and two-family residential developments to meet its "solar-oriented lot" definition, and development plans must protect access to sunshine for solar energy systems to the maximum amount feasible (§15.14.030.A.3). Dixon,

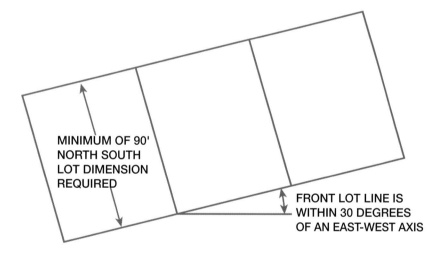

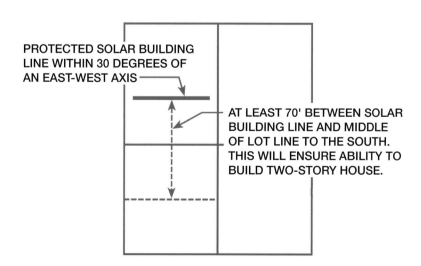

These figures from Clackamas County, Oregon's Zoning and Development Ordinance illustrate solar site design standards.

Source: Clackamas 2013

California, includes solar orientation and incorporation of solar energy systems in its list of general site design standards for single-family homes (§12.19.21).

Solar-Ready Requirements

"Solar-ready" provisions in local building codes require new construction to be electrically wired and plumbed to support the later installation of solar PV or hot water systems and may also require roofs to be oriented, designed, and built to easily accommodate and support those systems. Constructing a building to solar-ready standards is of little use, however, if the construction details are not available when the owner is ready to install a new system. Local governments have roles both in encouraging solar-ready construction and in holding solar-ready documentation in the building's permit history.

Some states, such as New Mexico, are adding solar-ready provisions to their energy codes. Local examples include Chula Vista, California, which has added PV pre-wiring requirements to its electrical code and solar water heater pre-plumbing requirements to its plumbing code. These require all new residential units to include electrical conduit and plumbing specifically designed to allow the later installation of solar energy or hot water systems (§15.24.065; §15.28.015). In a different approach, Henderson, Nevada, offers solar readiness as one of a number of sustainable site and building design options developers can choose in order to earn points required for development approval (§19.7.12).

Solar Mandates

A few communities have taken solar-ready building requirements to a logical extreme by mandating the installation of solar energy systems with some types of new development. In essence, these communities condition approval of development on either the provision of on-site PV systems or a financial contribution equal to the cost of purchasing the same amount of solar power.

Aspen-Pitkin County, Colorado, adopted the first similar type of mandate in 2000, though it was not specific to PV systems. The county added building code provisions that require homes over 5,000 square feet as well as certain energy-intensive accessory features (i.e., pools, spas, and snowmelt systems) to either install on-site renewable energy systems such as PV, solar thermal, or geothermal systems or pay mitigation fees (§11.32).

More recently, Lancaster and Sebastopol, California, both added provisions that require PV systems with new residential development. Lancaster's code requires residential developers to either install a specific amount of on-site PV

Lancaster, California, is the first U.S. city to require developers to install solar energy systems with every new home they build.

Lancaster Power Authority

capacity based on the number of dwelling units and the zoning designation of the site or to purchase an equal capacity of solar energy credits from an off-site installation (§17.08.060; §17.08.305). Sebastopol mandates PV systems for all new residential and commercial development, and it bases required capacity on building square footage rather than the number of units (§15.72).

Given the small number of examples, it is difficult to predict whether or not solar mandates will become a common strategy for growing local solar markets. In any given community, success depends on high development demand, strong local support for solar energy use, and a legal framework that supports development exactions.

DEVELOPMENT INCENTIVES

The goal of development incentives is to entice developers to provide a public benefit that they would not otherwise provide in exchange for increases in development potential, streamlined approval processes, or lower development costs. Development incentives for solar energy systems include any flexible application of zoning standards such as a density bonus or a height allowance, and any preferential treatment in the permitting process, such as discounted permitting fees or expedited reviews that favor solar over nonsolar development.

When considering new incentives, planners should consider both the actual value of an incentive to an owner or developer and the ability of the local government to efficiently deliver that incentive. Poorly planned incentives may detract from other more effective strategies for promoting solar energy use. Furthermore, if an incentive is too complex to understand or apply, the developer will opt for the familiar "regular" development process, leaving the incentive on the table. Often, the most effective approach is to talk to local developers and solar energy advocates and then craft incentives based on a careful consideration of all the relevant local development factors, such as development fees, market conditions, other zoning or building code requirements, and staffing levels.

Flexible Development Standards

Providing a limited exception to specific district dimensional standards (as referenced above) is, perhaps, the simplest way to create a small incentive for solar development. However, the most direct way to incentivize solar energy systems through zoning is to provide landowners additional development potential. This may take different forms, but the most common method is to provide additional floor area or height in exchange for passive solar design or installing a solar system that offsets a certain amount of onsite energy use. For example, Portsmouth, Virginia, allows developers to earn an additional one or two stories of development if a project includes a system that provides 20 percent of the project's electrical needs, along with other significant green building features (§40.1-5.8).

With additional density, planners must take care in calibrating the numbers on both sides of the incentive to reflect local market conditions and to ensure the local government is not either giving away too much development for minimal solar gains or asking for so much that no developer can afford to take the deal. Also, it is particularly important to anticipate possible opposition from neighbors who might not be pleased with more development or larger buildings in their neighborhood. The heavy use of an incentive can also make some citizens feel that the base zoning requirements no longer apply and that they can no longer know what to expect in their neighborhoods. For these reasons, it may be useful to limit the availability of density bonuses to certain areas of the locality, such as downtowns, commercial corridors, and multifamily areas that can more easily absorb additional development.

LIMITING PRIVATE RESTRICTIONS

Private homeowners' association covenants or design review requirements that prohibit or restrict solar energy systems are relatively common barriers. Though these provisions typically fall outside of local government control, more than half of all states have passed solar rights laws that either limit the restrictions that private covenants can place on solar energy system installation or explicitly enable local governments to adopt regulations aimed at protecting solar access (DSIRE 2013c). Planners in these states can raise awareness around this issue and ensure that residents and local officials understand when private restrictions on solar energy systems will be preempted by state or local protections. In states without such protections, planners can encourage home-rule municipalities to adopt local provisions limiting private restrictions on solar energy systems.

Another effective incentive is to provide developer relief from certain development standards. For instance, given that parking is a major expense for most projects, allowing a reduction in the required parking in return for a certain level of solar development will save a developer considerable money to help pay for that solar energy system. Similarly, reducing landscaping requirements (especially shade trees) for buildings using solar energy can help offset the additional cost of passive design features or a solar energy system.

Because a developer's need for flexibility is often project specific, the best strategy may be to provide a menu of development standards than can be adjusted to respond to the needs of different projects. For example, if solar access is limited on a development site by existing structures, then allowing a proposed structure (not simply the solar collector) to encroach into a setback to maximize solar access or allowing it to be a little taller than otherwise permitted may be necessary to get a solar energy system integrated into that project.

Reducing Permitting Costs

Apart from substantive changes to zoning standards, a community can also make procedural changes that facilitate solar development. The most common approach is to streamline the development review process for projects that include solar energy systems. Most developers use debt financing, and the more quickly they can get a project approved, built, and sold, the more quickly they can repay their loans—which increases their profit margin.

Also, if the streamlined process is designed to be more predictable, then developers can better plan and finance their projects from the beginning, rather than forcing them to prepare for unpredictable review processes that frequently get mired in delays.

Techniques to streamline review procedures include the following:

- *Add a pre-application checklist and meeting:* A good way to avoid costly and frustrating delays in a development review process is for local staff to clearly convey to the developer early in the process what the likely development issues, procedures, and expectations will be for the proposed project. This establishes an open line of communication between the staff and developer and helps both sides work together to move the application through the process.

- *Allow online submittal of solar energy system applications:* A significant portion of a solar energy system's cost comes from the time and resources consumed by printing, organizing, and delivering multiple hard copies of one or more permit applications to the appropriate review agencies. By allowing such applications to be filed online, applicants will save these costs—with the added bonus that the review time is often reduced as well. Portland, Oregon, allows solar applications to be filed online and staff can usually complete its review within two days (Portland 2010).

- *Increase ministerial approvals:* Time-consuming public hearings before appointed or elected bodies are sometimes required for approvals that could be effectively reviewed and approved at the staff level. Therefore, as referenced above, adopting objective review standards and switching to ministerial reviews for most solar development scenarios can be an effective streamlining tool.

- *Consolidate permits or approvals:* Because solar energy systems interface with multiple building elements, they may require multiple permits, including electrical (for PV systems), roofing, mechanical, and plumbing (for hot-water systems). The need to get approval for multiple development permits can delay projects. Thus, consolidating the reviews of permits, such as development plan approval and building permit review for an accessory solar system, or creating a single, but separate, permit for PV and thermal systems can save considerable time. The Solar America Board for Codes and Standards' Expedited Permit Process for PV Systems is, perhaps, the most widely referenced resource for communities interested in streamlining structural permitting processes for small accessory PV systems (Brooks 2012).

- *Expedite application processing:* Applications for projects that incorporate solar development can be given preferred status and reviewed and acted upon by the jurisdiction more quickly than nonsolar applications. However, this strategy only acts as an incentive if the existing review process is relatively arduous, such that a quicker process would provide significant relief to the applicant. Also, some communities make the mistake of promising an expedited review but then do not have the staff or capability to effectively coordinate all the moving parts to make it a reality (e.g., ensuring that other local review departments adhere to the same accelerated deadline).

Finally, because application fees can constitute a significant part of a developer's total cost for a project, a reduction or waiver of these costs in exchange for incorporating passive solar design techniques or solar energy systems is one more way communities can encourage solar energy use. Most communities use a flat fee or base the application fee on the value of the solar system. Research shows that the valuation method tends to penalize landowners by creating higher fees than flat-fee methods because more expensive systems do not necessarily require additional review time (Mills and Newick 2011).

Many solar experts recommend a flat fee of no more than $300 for most solar PV systems, meaning that many communities might need to reduce their current fees. In practice, solar permit fees vary widely from $0 in some communities to as high as $1,200 in others (DSIRE 2013a).

▶ SAN JOSE, CALIFORNIA

San Jose, California, is a leader in solar development and has achieved this status in part by aggressively streamlining its review requirements for solar permits. In most jurisdictions, new solar systems may require building, planning, and electrical permits. However, San Jose requires no building plan review for solar systems that meet basic and common conditions, such as having a panel weight of less than five pounds per square foot, being flush-mounted on the roof, and not exceeding 18 inches in height (San Jose 2012). This exemption covers most residential solar systems and saves property owners considerable time and money. In addition, no planning review is required for solar systems on single-family or duplex dwellings. When planning permits are necessary, such as for some multifamily structures and commercial projects, they are provided over the counter (Mills and Newick 2011).

The city requires electrical plan review for multifamily, commercial, and industrial PV installations, but only requires this review for single-family and duplex installations applicants in select instances based on the complexity of the installation (San Jose 2012).

The time and cost for final inspection of the installed PV system is another procedural requirement that can be a significant barrier to solar development. San Jose, however, has trained its inspectors through workshops and other professional development opportunities to closely and quickly conduct their inspections, often using standardized checklists, so that inspections typically take less than an hour (Mills and Newick 2011). This saves time and money for the city, solar contractors, and ultimately solar customers.

FINANCIAL INCENTIVES

Financial incentives for solar development are offered by every level of government, as well as by utilities and nonprofits. However, consumers, and even local governments, may be unaware of these incentives or have difficulty taking advantage of them. Understanding the types of financial incentives available for solar development and how these programs are structured may help planners, public officials, and solar advocates make a better financial case for solar energy use.

Most financial incentives for solar development can be categorized based on two sets of attributes (Figure 5.1). First, an incentive may be classified either as an up-front incentive (UFI) that is redeemed at or near the completion of an installation, or as a performance-based incentive (PBI) that is awarded over time as an installation generates electricity. Second, an incentive may be classified as either cash- or tax-based. The combination of these traits defines the basic nature of an incentive. Other incentives that do not fall within these basic categories include those that may accrue over time without any relationship to project performance, such as accelerated depreciation and property tax exemptions or special assessments. Lastly, programs that provide direct financing options for consumers and businesses (e.g., a subsidized loan program) comprise a distinct category of incentive.

Up-Front and Performance-Based Incentives

As the name implies, UFIs are generally paid or redeemed in a lump sum at or near the time of installation. The term UFI is more commonly used to refer to cash-based incentives than tax-based incentives. Cash-based UFIs are commonly designed as rebate programs that provide an incentive to all applicants who meet detailed program eligibility criteria. However, they are sometimes structured as competitive grants, where the grantor weighs individual applications against each other on the basis of a scoring and evaluation system. Tax-based up-front incentives often take the form of investment tax credits or sales tax exemptions. Incentive amounts for UFIs are typically stated in terms of dollars per watt of capacity ($/W) or as a percentage of the installed cost.

In contrast, PBIs are based on the actual electricity produced by a PV system or displaced by a thermal system on a dollar per kilowatt-hour ($/kWh) basis, either in the form of cash or as a tax credit. Cash-based PBIs can be designed as standardized, fixed contracts between a system owner and another entity (e.g., a utility), or as market-based incentives that fluctuate

***Figure 5.1.** Basic categories for financial incentives*

American Planning Association

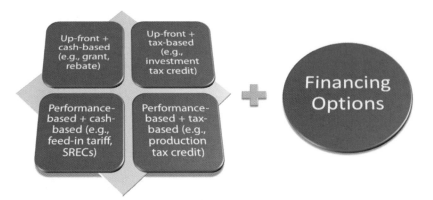

in value according to supply relative to the demand created by a state's renewable portfolio standard.

Market-based programs require the system owner to make additional arrangements to realize revenue from solar renewable energy certificate (SREC) sales, such as customized contracts with a utility or energy supplier. It should also be noted that some UFI programs require awardees to surrender the SRECs produced by a system in exchange for an incentive.

A standardized PBI program can be set up as a standalone program that does not involve the sale of electricity or SRECs, as an SREC purchase program, or as a bundled sale of electricity and SRECs. Standardized PBI programs typically have contracts that guarantee a certain incentive level over terms ranging from 5 to 20 years. Where a PBI involves a sale of electricity, the counter-party must generally be a utility.

The term feed-in tariff (FiT) is often used to describe programs that offer long-term guaranteed contracts for both electricity and SRECs. In contrast to programs that only involve an SREC purchase, under a FiT all of the electricity produced by a system is fed back into the utility grid and the system owner is not entitled to use any of the power on site (e.g., under a net metering arrangement). Under a standalone program, the system owner receives an incentive, but retains the ability to use the electricity on site and retains ownership of the associated SRECs. The basic advantages and disadvantages of UFIs and PBIs are outlined in Table 5.1.

Table 5.1. Up-front incentives versus performance-based incentives

Incentive Type	Advantages	Disadvantages	Examples
Up-Front Incentive	• Can be simple and easy to understand • Directly reduces up-front costs • Easy to budget one-time influxes of funding • Incentive levels can be designed based on expected performance of individual systems	• May suffer from cost-effectiveness concerns if performance verification is not addressed • Incentives based on expected system performance may require extensive and complex documentation, reducing program simplicity • Incentive amount is often not sufficient enough to affect the ultimate project feasibility	• New York State Energy Research and Development Authority: Small Customer-Sited PV Incentive Program (www.nyserda.ny.gov) • California Solar Initiative Expected Performance-Based Incentives for Small PV Systems (www.cpuc.ca.gov/PUC/energy/solar) • Oregon Department of Energy: Renewable Energy Development Competitive Grants Program (www.oregon.gov/energy)
Performance-Based Incentive	• Inherently rewards better-performing systems • Creates a further incentive for proper system maintenance • Guaranteed incentives facilitate financing	• Requires on-going reporting and verification, resulting in continued administrative costs • Guaranteed incentives necessitate dedicated funding over an extended period of time • Does not reduce up-front costs • Fully market-based SREC programs do not provide guaranteed revenue and require additional efforts on the part of the system owner to realize SREC value	• Maryland Public Service Commission: Market-Based SREC Program (http://webapp.psc.state.md.us/intranet/ElectricInfo/) • Delaware Sustainable Energy Utility: Long-Term SREC Contract Program (www.srecdelaware.com) • Gainesville Regional Utilities (Florida): Solar Feed-in Tariff (www.gru.com) • Washington: Renewable Cost Recovery Incentive Payments (http://dor.wa.gov)

One of the more recent variations in program design that blurs the line between UFIs and PBIs is the adoption of expected performance-based incentives. These programs provide an up-front incentive, but adjust it from a base level to account for individual system characteristics that affect energy production. These characteristics may include revised PV-panel nameplate ratings that account for real-world operating conditions, inverter efficiency, project location, and siting characteristics such as system orientation and potential shading. The best design for a particular incentive program varies with the goals and participants of the program (Barbose et al. 2006).

With any fixed incentive, one of the primary challenges a program administrator faces is setting the incentive level and adjusting it over time to reflect changing market conditions. Generally speaking, any program that is intended to last for more than a short period of time should allow for such adjustments. Some programs specify a standard adjustment methodology in advance by "stepping down" incentives as certain benchmarks are reached, such as the number of systems or amount of generating capacity enrolled in the program. Others simply provide for periodic (e.g., quarterly or annual) review by program administrators. Standardized adjustment schedules provide greater certainty for potential program participants, but are typically not as effective as customized adjustments at reflecting market changes.

Tax-Based Incentives and Cash-Based Incentives

Federal, state, and local governments have adopted a wide variety of tax-based incentives that address the primary modes of taxation in the U.S.: income, sales, and property taxes. Of these, income tax benefits are currently the most significant and broadly applicable federal incentives for solar energy systems (DSIRE 2013d). Table 5.2 shows major federal tax incentives for solar energy systems. Of all the incentives offered at the state level, income tax credits make the biggest impact in terms of reducing the cost of a system because of the magnitude of most income tax credits. Some states offer income tax credits as high as 30–35 percent of system cost, while sales and property taxes account for a much lower percentage of a project's costs.

Income tax credits are most often based on a percentage of the system's capital cost rather than on a dollar-per-watt basis. As with cash-based incentive programs, tax-based incentives may necessitate adjustment over time to avoid the over-subsidizing of solar projects as costs decline. In this respect, an income tax credit based on a percentage of installed system costs is somewhat self-adjusting because declines in capital costs also result in a decline in the amount of the tax credit without requiring frequent statutory changes.

Tax incentives are often differentiated by whether they can be applied against personal income taxes, against a variety of business-related taxes, or both. Tax credit laws commonly contain provisions establishing maximum

Table 5.2. Major federal tax incentives for solar

Incentive	Incentive Summary
Business Energy Investment Tax Credit	30% of installed costs for solar systems put to a business use
Modified Accelerated Cost Recovery System (MACRS) + Bonus Depreciation	Shortened depreciation schedule (5 years) for solar systems eligible for the Business Energy Investment Tax Credit. Through 2013, 50% first-year bonus depreciation
Residential Renewable Energy Tax Credit	30% of installed costs for solar systems that produce energy for use in a dwelling unit used as a residence by the taxpayer

incentives at different levels for different sectors and the carryover of excess tax credits to subsequent years. Some also contain aggregate limits for all tax credit claims in a year or multiple years, provisions addressing third-party owned systems, clauses that allow tax-exempt entities to benefit from the credit, and additional system equipment and design requirements. Tax credits intended to support residential installations are often limited to solar energy systems owned by the residential personal income taxpayer, though some states (e.g., Oregon and New York) now permit them to be claimed for residential systems owned by a third party (DSIRE 2013e).

Sales tax incentives are typically formulated as full exemptions from state sales taxes for solar energy systems, but they may also apply universally to local sales taxes or allow local governments an option to create a local exemption. Property tax incentives display greater variation, but they are most commonly structured as exemptions or special assessments that reduce the assessed value of a system from what it would otherwise be. In addition, state-level property tax incentives may also include local option provisions. Property assessment, taxation, and exemption laws for small behind-the-meter PV systems frequently differ from those applied to large-scale projects that generate electricity for sale (Barnes et al. 2012). Overall, property and sales tax incentives often make smaller impacts on project costs or ongoing operating expenses, but they can be significant enough to influence purchasing decisions in locations with high state and local tax rates. For instance, consumers in New York City pay a rate of more than 12 percent between state and city sales taxes, thus the sales tax exemption offered by the state and city for residential solar installations carries significant benefits (DSIRE 2013f).

Similarly, there are instances where property taxes have meaningful impacts on solar development. For example, under Ohio law, solar developers who sell power to third parties have been subject to personal property taxes as a public utility, which led to tax bills as high as $115,000 per megawatt. Starting in 2014, these projects are now eligible for a property tax exemption in exchange for payments in lieu of taxes that range from $6,000 to $8,000 per megawatt (Bricker and Eckler 2011).

Incentives that are not tied to a tax are considered cash-based incentives. Such incentives can include rebates, grants, or cash payments for RECs or electricity produced (i.e., PBIs). Rebate programs are common for relatively standard residential and commercial PV or thermal installations, while competitive grants may be targeted at larger PV installations, at a certain sector (e.g., schools), or with specific program goals that go beyond simply supporting new installations (e.g., educational opportunities). Rebates tend to have a greater overall market impact than grant programs because the plug-and-play nature typically provides more certainty for applicants and their contractors. In terms of PBIs, both REC programs and feed-in tariffs are considered cash-based incentives, while other PBIs can be designed as either tax- or cash-based incentives.

Because cash-based incentives are not related to taxes owed, cash-based incentives can be offered by a wide variety of entities, including governments, utilities, and even nonprofits. Cash incentives can be preferable for system owners because they are more flexible and do not require the applicant to have an existing tax liability, which also means that cash-based incentives can be made available to public and nonprofit organizations and low-income households. However, government entities often prefer tax-based programs over cash-based incentives because they do not require a dedicated source of funding. Both cash-based and tax-based incentives can prove to be important components of stimulating solar market development, but they each have benefits and drawbacks as outlined in Table 5.3 (p. 70).

Incentive Type	Advantages	Disadvantages	Examples
Cash-Based Incentive	• Cash is readily useable by any party • Well-suited for adoption of measures that support ongoing performance evaluation and verification • Tend to be more flexible, as market changes and emerging questions can often be addressed without changes in law • Can likely be offered by any entity if funding is available	• Require a ready and dedicated source of funding and represent an easily identifiable "cost" • Typically require a dedicated professional program staff and processing system • May actually result in additional tax liability if treated as income, effectively reducing the incentive	See up-front and performance-based incentive examples from Table 5.1.
Tax-Based Incentive	• Forgone revenue is less likely to be perceived as a cost, possibly rendering tax incentives more politically attractive • Can utilize existing processing and administrative infrastructures	• For income tax incentives the value is limited to parties with a tax liability, thus not directly usable by tax-exempts and less useful for taxpayers with minimal tax liabilities • Lack of dedicated and ongoing oversight may fail to promote maximum performance • Ambiguities and emerging issues with the tax incentive may be difficult to address without changes in law or a complex determination process • Navigating tax laws can be intimidating for consumers • Local programs may not be possible, or may require state authorization	• North Carolina: Renewable Energy Tax Income Tax Credit (www.dsireusa.org) • New York: Residential Solar Income Tax Credit (www.dsireusa.org) • Arizona: Renewable Energy Production Tax Credit (www.azdor.gov/TaxCredits/) • New York: Solar Sales Tax Exemption Local Option Exemption (www.dsireusa.org) • Ohio: Property Tax Exemption for Small Energy Facilities (www.dsireusa.org)

Table 5.3. Cash-based versus tax-based incentives

Loans and Similar Financing Options

Regardless of the other incentives available, a solar project will typically still carry significant up-front costs for the owner or purchaser. This remaining up-front cost can, of course, be paid directly out-of-pocket, but many purchasers cannot afford to pay cash for the balance, or would prefer to spread out the expense over time for other reasons. Furthermore, traditional financing options may prove infeasible or unattractive for a number of reasons. For instance, a homeowner may lack adequate equity for a home equity loan or fail to meet other underwriting standards, or solar may be considered risky by an unfamiliar lender, resulting in a high interest rate. Recognizing the impediments to traditional financing, some state and local governments offer loans with lower interest rates, longer amortization periods, or lower fees than loans available from commercial lenders (CESA 2009).

In practice, government-sponsored loans take the form of either direct loans or public-private partnerships. Many state governments offer direct loans for solar energy systems and other renewable energy and energy-efficiency projects through revolving loan funds, where loan payments are returned to the loan pool and made available to future borrowers. These programs require a significant initial source of capital, but if structured properly, can exist indefinitely. However, many revolving loan funds target only subdivisions of state and local governments or specific sectors such as agricultural businesses, and they are not available to consumers.

Rather than directly administer loan programs, some state and local governments have partnered with private banks in developing special financing programs specifically for consumer energy-efficiency and renewable-energy projects. In programs of this type, the governmental body is typically able to make a small commitment of capital to serve as a credit enhancement, resulting in more favorable terms for consumers. Credit enhancement can take a number of forms, such as interest rate buy-downs, loan guarantees, or the establishment of a loan loss reserve fund. Regardless of the model employed, the primary virtue of these public-private partnerships is greater scale, where a small financial commitment on the part of a governmental agency is able to leverage a much larger amount of private capital.

Government-sponsored loan programs are challenged by the necessity of raising an initial pool of loan capital and by the fact that loan repayment terms may exceed the length of time that a property owner expects to remain in a building or home. In recent years, some state and local governments have begun exploring a novel option for providing consumer financing options for energy improvements that addresses both of these issues. Under so-called Property Assessed Clean Energy (PACE) programs, a unit of government (usually a local government) loans money to a property owner for clean energy improvements, and this loan is then repaid via a special assessment on the property tax bill. Currently 29 states and the District of Columbia have authorized PACE programs (DSIRE 2013g).

If a property participating in a PACE program is sold, the assessment can remain attached to the property, and the new owner will continue paying the assessment. Programs are designed so that the added assessment is less than or equal to the energy savings realized by the project, resulting in savings for both the original and new owners. The assessment is considered a lien on the property, which in theory is secure enough to permit the local government to borrow money to capitalize the program at rates low enough to permit an attractive loan offering to consumers. While residential PACE programs have been stymied by concerns raised by the Federal Housing Finance Agency over these lien provisions, a number of commercial-sector programs have successfully gone forward (PACE Now 2013).

Possibly in recognition of the uncertainty surrounding the future of residential PACE, some jurisdictions have begun to investigate and implement utility on-bill financing programs for energy improvements. As with PACE, on-bill financing ties the repayment mechanism to the property (by way of the utility account). It also takes advantage of the ability of many utilities to raise large amounts of low-cost capital through existing channels or through collections from ratepayers. Utility on-bill financing, particularly as applied to solar projects, remains in its infancy, but a statewide program is currently being developed in Hawaii (Hawaii Public Utilities Commission 2013).

▶ AUSTIN, TEXAS

While the state of Texas has done little to incentivize solar development, its capital, Austin, has a long history of supporting solar energy use and more than a decade of experience with incentive programs. Having its own municipal utility, Austin Energy, gives the city access to ratepayer funds for incentive programs and greater flexibility in designing rate structures. Austin Energy's current portfolio of PV programs includes several types of incentives, each designed to address a specific barrier to the various market segments. Commercial PV systems up to 200 kW are eligible for a PBI (sale of SRECs) of $0.14 per kilowatt-hour (kWh) and commercial systems of 20 kW or less remain eligible for net metering. For homeowners, the utility offers a UFI of $2.00/W and a consumer solar loan program (Austin Energy 2013). Austin Energy also recently replaced residential net metering with its Value of Solar Tariff, which currently credits electricity production from systems of 20 kW or less at a rate $0.128/kWh (Austin Energy 2013). These incentives have also encouraged the nearby city of Sunset Valley, which is served by Austin Energy, to supplement the utility rebate with its own citywide rebate program for residential PV. The Sunset Valley program essentially piggy-backs on the program infrastructure established by Austin Energy, allowing it to avoid many of the burdens involved in designing and hosting a local program (Sunset Valley 2012).

▶ BOULDER, COLORADO

In 2006, Boulder, Colorado, passed the Solar Rebate Ordinance, which established a solar sales and use tax rebate for PV and solar water heating installations (§3-2-17.h). Owners of solar systems may receive a rebate (essentially a sales tax refund), which is drawn from the city's Renewable Energy Fund (REF). The REF is a portion of the city's general unrestricted sales and use taxes collected on eligible solar system installations. The fund is used to provide the sales tax rebate (approximately 15 percent of the city sales tax paid on a system), and the city's Solar Grant Program, which provides grants for local nonprofit organizations and homeowners in the city's affordable housing program. The grants vary case by case but are limited to 50 percent of the cost of the system (Boulder 2013).

SUMMARY

Collectively, regulations and incentives represent the third strategic point of intervention for communities looking to promote solar energy use through planning. When local zoning, subdivision, or building codes fail to explicitly sanction different types of solar development, it can create uncertainty for property owners and other stakeholders interested in solar energy use. Consequently, clear definitions, use permissions, and development standards for different types of solar development are fundamental to a solar-supportive regulatory environment. Furthermore, both development and financial incentives can play significant roles in making solar development feasible.

CHAPTER 6

Development Work

*Darcie White, AICP, Paul Anthony, AICP, Brian Ross,
and David Morley, AICP*

 The fourth strategic point of intervention for communities looking to promote local solar energy use through planning is development work. Since most local plans depend heavily on private investment for successful implementation, it is important for localities to consider the roles that review and participation in land development can play in promoting passive solar design and solar energy systems. For the purposes of this report, development work includes activities that commonly fall under the umbrella of development services as well as public-private development partnerships.

This chapter begins with a discussion of how the two key aspects of development services—permitting assistance and development review—intersect with efforts to encourage solar energy use. Subsequent sections provide an overview of various types of public-private partnerships that communities may use to actively participate in solar development projects.

DEVELOPMENT SERVICES

Development services include both the "over-the-counter" permitting assistance provided by the planning department to the general public, landowners, and developers and the more involved service of reviewing development applications for compliance with zoning standards. The two services are closely related, and communities should make every effort to carefully coordinate them to create the most efficient and seamless development services department possible. For instance, the better the front counter staff are at providing accurate, timely, and relevant zoning and permitting information for solar development, the better prepared applicants and the public will be to successfully navigate the development review process.

The goal is to establish clear expectations regarding what the development review process entails for different types of solar development so that costly and time-consuming surprises are avoided later in the process. In many localities, permit counter interactions between staff planners and home or business owners and development project review meetings may be the most visible parts of the local planning system.

Permitting Assistance

One area in which many planning and development departments can improve their performance is their over-the-counter service. Planners are often so busy reviewing development applications for compliance, preparing staff reports, addressing neighborhood concerns, and completing other duties that dedicating enough time and energy to create an effective front counter presence is difficult. This is unfortunate because providing the public with an array of informational and educational tools that can be accessed in the front counter area—both through speaking to a planner or by using web-based data, maps, and programs—empowers the members of the public by

The Solar Boston Permitting Guide describes the city's streamlined permitting process for PV installations.

Source: Boston 2010

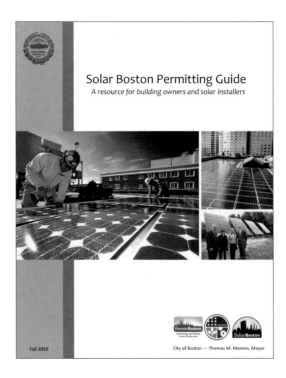

helping them answer many of their own questions. There are a number of strategies for maximizing the front-counter area in promoting solar energy use in the community.

Many planning offices already provide a variety of informational guides, brochures, or pamphlets that address common zoning and permitting requirements. For communities that are committed to promoting solar energy use, a solar energy pamphlet can be a great outreach tool. These materials should be concise, written for a general audience, and should contain photographs or illustrations. The goal is to not only summarize the community's solar policies and regulations, but also to explain the general benefits of solar energy systems and their critical role in our energy future. For example, Santa Barbara, California, has a Solar Access Packet that provides important regulatory information on solar access height limits, applicable code sections, and instructions for preparing a Solar Access Shadow Plan (Santa Barbara 2012). Sacramento, California, also has a very informative solar application guide (Sacramento 2011).

While printed pamphlets are useful, especially for less technologically inclined people, posting information online or adding public-access computers in the front counter area can greatly increase the amount and sophistication of solar-related information that can be delivered to the public. For example, Tucson, Arizona, advertises and provides convenient access to free webinars on solar financing, permitting, and other relevant topics that cover the latest best practices on the installation of solar systems (Tucson 2013). Another option is to develop a "solar tutorial" that not only takes the user through the basic parameters for solar systems, but shows photos and examples of how solar is being used locally, provides answers to frequently asked questions, and provides links to other helpful solar resources. Setting up such a tutorial requires an upfront investment in staff time, but it should also decrease the number of time-consuming inquiries that planning staff have to field over the long term. The Mayor's Office of Sustainability in Philadelphia offers a series of five tutorials on its website that walk prospective applicants through the basics of solar development, the reasons why solar development makes sense, different financial models for solar development, and the process of finding a solar installer (Philadelphia 2013).

Communities that want to take customer service to the highest level can establish a solar ombudsman. An ombudsman is typically a person established by a governmental entity to serve as an advocate for the public by investigating complaints and attempting to resolve them, often playing the role of mediator between the government and the public. An ombudsman will also commonly provide periodic recommendations for improving the efficiency of the involved governmental agency. In addition to the above roles, a solar ombudsman could serve as the point person for all development issues related to solar installations in the community. This would create a clear source for developers and others to get consistent and authoritative answers about solar-related policies, regulations, incentives, and resources. A solar ombudsman also need not be a full-time government employee but can be a solar expert from another entity. New York City, for example, has a solar ombudsman who is employed by the City University of New York but who works two days per week at the city's permitting department as a solar advocate and who assists the public with solar questions and guidance on the permit process.

As mentioned in Chapter 5, pre-application conferences are an effective way for staff to get the landowner, the landowner's agents, and other relevant governmental agencies in one room to identify and discuss the major issues involving a proposed development project. Given the more narrow focus of solar energy systems, a special pre-application meeting could be established to address only the issues pertinent to a proposed solar energy system, most likely with fewer planning staff than necessary for a general pre-application conference. This will provide another opportunity for the planning staff to educate the applicant and set realistic expectations early in the review process regarding development requirements and potential neighborhood concerns.

Development Review

The development review process is another opportunity for planners to directly influence the final outcome of a project and encourage solar energy use. However, planners are often too busy, inexperienced, or cautious (perhaps because they have not been explicitly empowered by superiors) to proactively advocate creative ways to integrate solar energy systems into proposed projects. Therefore, to capitalize on this opportunity, planning directors need to train staff members and give them the confidence to do more than merely check each development application for compliance with minimum development requirements. Because this added effort is not appropriate in every situation, it is also important for directors to set thresholds (e.g., based on project size or the anticipated energy load on-site). Like many sustainable development options, solar development is not the standard or easiest choice for many developers, so progress may depend on planners taking the initiative to become problem solvers and creative site designers to find ways to make solar energy use a reality in the projects they are responsible for reviewing.

The first rule of being an effective advocate for solar energy use during the development review process is to understand the project objectives from the developer's perspective. This is easier said than done for many planners. This is because many planners have little or no private-sector development experience and so are not very familiar with the financial and practical realities that face developers. Too often planners simply assume that emphatically stating the existence of a requirement or incentive is sufficient to shift the onus to the developer to find a way to comply. Left without further direction, most developers will simply find the cheapest way to comply with the minimum requirements, even though they are often willing to listen to alternative approaches if the jurisdiction is willing to provide some flexibility or incentives to create a better outcome—especially if that outcome would add long-term value to their project (see the discussion of development incentives in Chapter 5). Thus, planners need to anticipate the concerns of developers and advocate for projects that integrate passive solar design or solar energy systems and benefit both the public and the developer.

With this said, planners that advocate for solar development must also anticipate and balance the concerns of neighbors and other interested individuals or organizations, including decision makers who may be skeptical of solar energy systems or concerned about their impacts. The effectiveness of planners can often be measured by how well they are able to anticipate potential controversies and navigate a project through the political process without getting bogged down by NIMBYism, misinformation, or other obstacles. Planners also must make sure that they do not become blind advocates for solar energy use such that they are unable to see clearly the legitimate concerns neighbors may have about the glare, location, or aesthetics of solar systems and address those concerns fairly and transparently.

Finally, communities might want to consider using an outside solar expert to assist them during development review processes when they feel they lack the experience to adequately review proposed projects that include complex or controversial solar energy installations. Ideally, this expert would be from within the community to add credibility, but any qualified expert could provide significant assistance in the right circumstances (provided this individual is not a project competitor and does not have a vested interest in the outcome).

PUBLIC-PRIVATE PARTNERSHIPS

In order to achieve public goals, cities and counties often guide private development using tools such as education, incentives, and development regulation. In some instances, however, a local government can also be a financial partner in a development or redevelopment effort.

While most cities and counties own land in order to house specific public facilities, this is not the only reason localities own land. For example, local governments may also acquire land inadvertently, such as through tax forfeiture, or deliberately, such as when assembling parcels for redevelopment. Furthermore, as facility needs change over time, public entities may find themselves in possession of surplus properties. In most of these cases, the community's ownership is temporary, and the explicit purpose is to facilitate private development and achieve community benefits through the redevelopment process.

There is a long history of public-private partnerships (PPPs) to achieve development and redevelopment goals. Local governments acquire and assemble parcels, provide or upgrade infrastructure, demolish buildings, remediate brownfields, or make other investments to enable private-sector redevelopment. In some instances, local governments coordinate or lead public-private development projects that combine public and private investments on a single site. In all of these instances, the community can use its financial participation to encourage solar development and capture the public benefits of using local energy resources.

Public entities that engage in PPPs are generally not doing so, primarily, for financial gain. The purpose of public-sector participation is to capture the public benefits of redevelopment, such as minimizing the blighting effect of underused or vacant parcels and buildings, or to create job opportunities for community residents. By considering opportunities for solar development as part of the larger development process in PPPs, communities can enhance public benefits and help transition local energy markets to self-sustaining status.

PPPs can take a number of forms, but there are three types of partnerships that have special relevance for communities committed to promoting solar energy use:

1. PPPs involving the disposition or redevelopment of publicly owned land

2. PPPs involving joint development or co-development on publicly owned land

3. PPPs involving public financial support for development on privately owned land

Land Disposition and Redevelopment

Local governments have a range of options for dictating how vacant or undeveloped public lands are redeveloped. A hierarchy of options stretches from strategies that leave redevelopment entirely in the hands of the private-sector market to strategies in which the community sets very specific conditions on redevelopment activities. Communities can use any of these models to facilitate solar development. There are three scenarios that span this hierarchy:

1. Selling property on the open market, which can include a stipulation that development incorporates solar development.

2. Issuing a request for proposals (RFP), which describes the public (solar development) and private (market-driven development) purposes without providing specific designs or specifications, for a private entity to buy and develop a parcel.

3. Developing a specific redevelopment plan that includes solar development prior to soliciting bids, including specific designs that ensure public benefits, and requiring completion of the plan as a contingency of being awarded the redevelopment opportunity through an RFP.

The community's ability to require solar development as part of public land disposition and redevelopment will depend heavily on a number of local conditions. Market demand for development, characteristics of the site, and the community's long-term plan for the neighborhood or surrounding land all affect the efficacy of the PPP strategy. While incorporating solar development into land disposition actions is still rare, a number of communities have adopted land disposition policies to guide the sale of surplus public property.

One example is the thousands of acres of formerly privately owned land that is now in public ownership in Detroit. The Detroit Works project created a hierarchy with specific criteria for when land should simply go on the market, when it should be repurposed for a largely public purpose (such as green or blue infrastructure), and when other redevelopment characteristics can be required as part of the redevelopment process (Detroit 2013).

Communities' best opportunity to incorporate solar development into land disposition is when issuing an RFP (rather than simply listing property for sale). An RFP stipulates the outcomes upon which the disposition of the property will be based. Communities should recognize that RFPs for development or redevelopment are most effective in a robust development market. In a less robust market the community may also need to provide the capital, or other subsidy, in order to ensure the project moves forward. In some cases, the community plays the role of the developer, assuming full risk for the development, but also being in the best position to ensure that public benefits and goals are included.

Communities can create a concept plan for a redevelopment area to identify specific public benefits to be achieved through redevelopment, similar to a small area plan process. When the area is publicly owned and market conditions are robust, the community's plan can become the development RFP; the community is a direct financial participant in the development, and thus has far greater leverage than relying on development regulation and incentives.

Eden Prairie, Minnesota, is following such a process in developing an undeveloped publicly owned 8-acre parcel into a "green" neighborhood development. The RFP requires that bidders, on acquiring the land for development, propose green elements from a menu of options based on the LEED Neighborhood Development standards, including an element for energy production and distribution. The RFP does not dictate specific designs or technologies, but it does require that proposers submit concept designs demonstrating how the green elements are being met (Eden Prairie 2013).

Communities can use an RFP process to include solar development in a number of ways. For instance, the RFP can include a requirement for passive solar design features or solar-ready buildings. A more definitive process is to require solar energy systems to be incorporated into buildings. A number of housing authorities have included solar systems as either a requirement or an alternative for meeting green building goals. In 2012 the Denver Housing Authority entered into a public-private partnership to install 2.5 MW of solar energy on more than 387 affordable housing buildings (Proctor 2012).

Finally, communities can include solar development in RFPs for public infrastructure in cases where the infrastructure has some synergy with solar development (see next section for joint-development examples with transit). Several communities have incorporated passive solar design features or solar energy systems in transit station projects, using public-private financing arrangements.

Joint or Co-Development

Parks, streets and rights-of-way, and public buildings all serve public purposes and are generally not privately owned or developed. But, in some instances, communities may elect to integrate private-sector activities on public sites, using joint development or co-development to provide services or capture value that would otherwise be lost. In these cases, planners may have opportunities to encourage solar energy use.

Joint development is a concept used in transit planning to mean the common use of property for both transit and nontransit purposes. The Federal Transit Administration (FTA) has specific guidelines for joint development that allows the use of FTA funds for a variety of nontransit activities, including renovation of historic transit facilities, development of facilities that house community services, and support of greenhouse gas reduction strategies that are synergistic with transit facilities (FTA 2013).

A similar concept, co-development, describes a PPP formed to coordinate the development of transit infrastructure (usually a transit station) with adjacent private development. Since the transit facility adds value to the private development, the private-sector partner may share infrastructure costs with the transit agency.

Communities can promote solar energy use through joint or co-development in a number of ways:

- Requiring all private development on transit sites to incorporate passive solar design features or to generate a set percentage of estimated energy use via on-site solar energy systems.

- Dedicating space for a privately developed solar PV system on the transit site in exchange for a power purchase agreement to sell power back to the transit authority or a lease payment to the transit authority.

- Incorporating solar development into housing or mixed use development adjacent to the transit station.

The Denver Housing Authority's Benedict Park Place 5B development has a building-integrated PV system that offsets common space electricity needs as well as 15 percent of unit power consumption.

Denver Housing Authority

Several recent or planned transit investments involve joint or co-development that incorporates solar development. For example, the plan for the Fenway Center in Boston incorporates a number of green elements, including a rooftop solar energy system intended to offset all power usage of the associated commuter rail station (Meredith Management 2011). A smaller example of a transit-oriented development project is the Patton Park Apartments in Portland, Oregon. Portland's transit agency, TriMet, initiated this co-development by purchasing land near a light rail station and issuing a request for qualifications (RFQ) from developers to build transit-oriented affordable housing. TriMet's RFQ required compliance with the Portland Development Commission's green affordable housing guidelines, which stipulate solar-oriented site design and passive solar building design features (TriMet 2006). The completed project meets these basic thresholds, and its owner is considering installing a rooftop solar system to offset on-site electricity usage (REACH n.d.).

Financial Support for Private Redevelopment

Public-private partnerships can also be limited to financial relationships rather than physically located on public land or infrastructure. Financial PPPs are frequently used to meet economic development goals, such as cases where the community enables a private-sector project to move forward when it might not otherwise have done so through mitigation of financial risk or market failure. These PPPs also offer an opportunity to enable solar development that would not have otherwise happened.

Whenever the community is a financial partner in a redevelopment effort, the community has opportunities to ensure that public benefits or amenities are included in the redevelopment. Financial partnering takes a number of forms, including direct subsidies and property tax abatement; tax increment financing, bonding, or loan guarantees that lower risk or interest rates; brownfield remediation or risk management; provision of infrastructure at public expense; and myriad other partnering options. Becoming a financial partner in the redevelopment process enables the community to go beyond the development standard requirements, including finding innovative ways to incorporate solar development within the redevelopment process.

The PPP tool can be used to directly support solar development, just as it would other types of development, or it can be used to leverage solar development within another development initiative. As an example of the former, Rockford, Illinois, used tax increment financing in 2012 for at least two solar projects: installing solar energy on the Bell School Reservoir and completing construction on a 3.5 MW solar farm (Rockford 2013).

SUMMARY

Development work is the fourth strategic point of intervention for communities looking to promote solar energy use through planning. Most local plans depend, at least in part, on private investment for implementation. Both development services and public-private partnerships provide opportunities for communities to remove barriers and provide incentives for solar development. Through development services communities can establish clear expectations regarding what the development review process entails for different types of solar development. And in some instances, communities may be able to leverage the benefits of public-private partnerships to incentivize private solar development.

CHAPTER 7

Public Investments

Philip Haddix, Chad Laurent, Jayson Uppal, Erin Musiol, AICP,
and David Morley, AICP

 The fifth and final strategic point of intervention for communities looking to promote solar energy use through planning is public investment. While private-sector solar development is likely key to substantial expansion of passive solar design and solar energy systems, public investments provide a chance for localities to lead by example. For the purposes of this report, public investment includes direct capital investments in solar development, third-party solar development projects where a public entity plays host to a solar energy system, and direct investments in economic development or educational programs. This chapter begins with a discussion of solar development on public facilities and concludes with a brief overview of how communities can support solar market growth through programmatic investments.

SOLAR ON PUBLIC FACILITIES

Public facilities require a lot of energy, both for electricity and space and water heating. With energy expenditures constituting about 10 percent of local government operating budgets, this demand can create a powerful incentive to reduce energy costs through on-site renewable energy production (USEPA 2011). Furthermore, public-sector energy demand also provides an opportunity for cities and counties to lead by example when it comes to promoting solar energy use. Vacant roof space on public buildings and idle or underutilized public land often hold potential for solar development. This section provides an overview of the types of solar projects that communities may pursue on public land and facilities, pointing out key considerations local governments can use as starting points for their own solar procurement efforts.

Solar Energy Systems on Municipal Buildings and Grounds

Local government buildings and the grounds upon which they are located present opportunities for reducing energy demands through passive solar design or offsetting those demands with solar energy systems. Before acting on an interest in solar development, it is important that communities understand how specific site characteristics affect each property's suitability for hosting a solar energy system. Consequently, communities will need to answer the following questions:

Who will host the system? Determining where the energy will be used will inform the system location selection process. Roofs providing ample space with few obstructions that are able to support the added weight of a solar energy system (between 2.5 and 6 pounds per square foot) and withstand wind loads may be good candidates for rooftop installations (Lisell et al. 2009). Such roofs should be structurally sound and not require replacement in the next 15 years (USEPA and NREL 2012). Properties with tracts of open ground space (e.g., schools and recreational facilities) that are not too steeply sloped (less than 10 percent grade) or can be graded cost effectively could serve as hosts for ground-mounted systems (USEPA and NREL 2012).

What type of energy is being used? As mentioned in Chapter 2, solar energy systems harness power from the sun in different ways and for different applications. Sites with a sufficient electricity load might desire to offset the amount of energy they obtain from conventional sources with a solar PV system. Conversely, properties with large hot-water demands (e.g., fire stations) might consider a solar thermal system.

Minnesota's Department of Natural Resources has installed a 13.8 kW roof-mounted photovoltaic system on a picnic shelter in a new campground at Lake Shetek State Park.

Minnesota Department of Natural Resources

Park City, Utah, installed a PV system on its city hall building in 2010. This system provides enough electricity for three average Utah homes and saves the energy equivalent of burning 23,000 pounds of coal each year.

Matt Abbott

What is the user's energy load? Regardless of the choice of technology or system location, contractors will need to assess a site's energy consumption in order to determine the appropriate size for a solar installation. A behind-the-meter PV system (i.e., one connected to the electric grid) will typically be sized to meet demand less than or equal to net site load requirements (Brooks and Dunlop 2012). In some states, net metering laws explicitly prohibit oversizing a system with respect to energy load, but even when oversizing is permissible, it can entail substantial additional costs that disproportionately exceed any added benefits. This principle of basing system size on energy demand applies to solar thermal technologies as well (Marken and Woodruff 2012).

What is the user's energy cost? The financial benefit of a system designed to offset on-site energy consumption is primarily driven by the rates end users pay for utility services. While a "back of the envelope" analysis may be useful as an initial step, each community will ultimately need to undertake a thorough analysis that considers both existing rates and other rate options in order to determine if a solar energy system will be an economically sound investment.

How will site characteristics impact performance? Because a solar system's energy output depends on the amount of sunlight it receives, a number of site-specific factors affecting where and how a system can be placed must be considered to optimize performance. Such site characteristics include, but are not limited to, system size, shading, array orientation, and module tilt.

Publicly owned sites that pass the prescreening and formal assessment processes and have sufficient electric or thermal energy loads can be strong candidates

A solar-powered parking meter on Ann Arbor, Michigan's Main Street

Dwight Burdette / Creative Commons 3.0

for solar installations. Examples of local government properties that have been fitted with solar devices designed to meet on-site demand include the following:

- City halls
- Agency offices
- Libraries
- Recreation centers
- Courthouses
- Public parks
- Community centers

- Correctional facilities
- Public safety buildings
- Educational facilities
- Health centers
- Parking lots and structures
- Street lights
- Parking meters

 MANSFIELD, CONNECTICUT

Mansfield, Connecticut, serves as a case study illustrating both the variety of properties that can be developed for solar PV and the economic benefits of such investments. Between 2008 and 2010, the town installed an 83 kW system on its community center through a power purchase agreement (PPA), as well as smaller arrays on its library (4.5 kW), senior center (8 kW), and two fire stations (4.5 kW and 6 kW). The town expects for these systems to satisfy a third of on-site demand at a net financial benefit to the municipality (Mansfield 2010).

Using the National Renewable Energy Laboratory's System Advisor Model (SAM), it is possible to estimate the financial aspects of systems similar to those installed in Mansfield (NREL 2013c). At current average installation prices, an 83 kW system such as the one installed at the community center would incur an up-front cost of approximately $350,000, had the city opted for direct ownership (SEIA and GTM Research 2013). With current Connecticut state incentives, this brings the levelized cost of electricity delivered by the system to 12.43 cents per kilowatt-hour (kWh), well below the latest state average of 14.56 cents per kWh (USEIA 2013). The system payback period is just over 13 years, with the investment netting the town over $26,000 in benefits over 25 years (a 7.4 percent return on investment).

A rooftop PV system installed on the Mansfield Community Center in Mansfield, Connecticut

Connecticut Clean Energy Finance and Investment Authority

MADISON, WISCONSIN

Many local governments have also had positive experiences installing solar water heating systems on municipal properties. Madison, Wisconsin, took great advantage of both solar thermal technology and available installation incentives between 2006 and 2008, when it installed solar water heating systems on 11 municipal fire stations (Focus on Energy 2008). These systems ranged in size from 96 square feet to 240 square feet (Madison 2013). The largest of these systems provides 60 percent of the site's hot water load (just under 600 therms annually), and state incentives reduced its up-front cost of $33,480 by nearly 25 percent. Through energy savings, the system will pay for itself in 19 years, ultimately yielding an 8 percent return on investment (Focus on Energy 2008).

Primary-Use Solar Energy Systems on Public Land

Beyond on-site usage for public facilities, idle or underutilized public land presents opportunities for local governments to develop or host solar gardens or farms.

In assessing the suitability of these lands for solar development, municipalities and contractors will need to answer many of the same questions that apply for systems serving on-site loads: Is the ground slope conducive to PV development? Is the property mostly free of obstacles and shading during peak solar hours? What restrictions do site characteristics impose on system size, orientation, and tilt? In addition to these considerations, factors such as distance to transmission and transportation infrastructure, the presence and nature of any land-use restrictions, and the opportunity cost to the local government of developing the land for solar development will be important in assessing site suitability (USEPA and NREL 2012). Many local governments have developed or hosted solar gardens or farms on sites such as the following:

- Landfills
- Water treatment facilities
- Brownfields

- Water reservoirs
- Municipal airports
- Surplus property

However, it is important to note that any site with documented or potential environmental contamination will require additional evaluation. For landfills, system siting and design will be influenced by characteristics including closure status, whether the facility has been capped and lined, soil traits (such as stability, settlement, and erosion), leachate and landfill gas management practices, and the existence of institutional controls or restrictions on redevelopment (USEPA and NREL 2012, 2013). Brownfields—lands overlooked for redevelopment due to real or perceived past contamination—entail their own unique considerations. Before solar development can occur on these properties, sites should be assessed for contamination and remediated (if necessary) and outstanding clean-up liability issues should be resolved (USEPA and NREL 2012).

Understanding available alternatives for the use of the electricity produced by a primary-use solar energy system is another crucial detail influencing whether and how solar development can and should be pursued on public lands. Solar gardens or farms can be designed to export power to the grid for resale to utility customers (the "grid-supply" scenario), offset the energy use of nearby public facilities (the "virtual net metering" scenario), or add to the generation portfolio of a municipal utility.

A 2.3 MW PV system on a capped landfill in Easthampton, Massachusetts

Borrego Solar Systems, Inc.

The output of solar gardens and farms often far exceed the amount of on-site electricity demand. In some cases, there will be no on-site load at all for the system to offset. In the "virtual net metering" scenario, a local government will still be able to benefit from the electricity the system produces if the state government has authorized virtual net metering, a policy allowing for net metering credits (see Chapter 2) to be allocated to multiple accounts. This policy can allow a local government to offset electric demand at several facilities, ensuring the community is able to reap the full benefits of the electricity produced by the system. Solar gardens or farms serving this purpose are either owned by a third-party (where state law permits) or by the local government itself.

The city of Easthampton, Massachusetts, had great success in redeveloping its closed landfill with a solar installation. The city benefits from state laws that both allow for third-party ownership and virtual net metering. Through these policies, Easthampton was able to install solar with no up-front capital investment on its part and purchase enough cheap solar electricity (at 6 cents per kWh) to power 20 percent of its municipal buildings. Over a 10 year PPA, the city expects to save over $1.4 million in electricity costs (Borrego Solar 2013).

In a "grid-supply" scenario, the local government merely serves as a host for the system. A solar contractor assumes most of the risk associated with the

The Jefferson County Courthouse in Golden, Colorado, incorporates passive solar design features in order to maximize the use of natural lighting.

David Parsons (NREL 0422)

project and retains responsibility for designing, financing, constructing, and maintaining the installation. Local governments do not directly benefit from the electricity produced by the system but instead receive lease payments from the system owner in exchange for the use of the property upon which the installation is sited (similar to the joint development scenario discussed in Chapter 6). In July 2012, the Indianapolis Airport Authority (IAA) realized firsthand the benefits of leasing land for solar development, approving a 30-year agreement to lease 75 acres at the Indianapolis International Airport for a utility-scale solar farm. For hosting the system, IAA will receive $315,000 annually from the developer, who will sell the electricity produced by the facility to the local utility (Sickle 2012).

PROGRAMS THAT SUPPORT SOLAR ENERGY USE

Apart from investing in passive solar design features or solar energy systems on public buildings and grounds, communities committed to promoting solar energy use also have opportunities to make programmatic investments that support solar development. These investments may fund mapping efforts that build awareness about solar energy use, or they may provide training and education programs designed to meet the solar industry's need for a skilled workforce.

COMMUNITIES WITH SOLAR MAPPING TOOLS

Anaheim, California: Anaheim Solar Map
http://anaheim.solarmap.org/

Berkeley, California: Berkeley SolarMap
http://berkeley.solarmap.org/solarmap_v4.html

Boston, Massachusetts: Solar Boston
http://gis.cityofboston.gov/solarboston/

Cambridge, Massachusetts: Solar Tool v.2.
http://cambridgema.gov/solar/

Denver, Colorado: Denver Regional Solar Map
http://solarmap.drcog.org/

Los Angeles County, California: Los Angeles County Solar Map
http://solarmap.lacounty.gov/

Madison, Wisconsin: Solar Energy Project (MadiSUN)
http://solarmap.cityofmadison.com/madisun/

Milwaukee, Wisconsin: Milwaukee Solar Map
http://city.milwaukee.gov/milwaukeeshines/Map.htm

New Orleans, Louisiana: New Orleans Solar Calculator
http://neworleanssolarmap.org/

New York, New York: New York City Solar Map
http://nycsolarmap.com/

Orlando, Florida: Metro Orlando Solar Map
http://gis.ouc.com/solarmap/index.html

Portland, Oregon: Oregon Clean Energy Map (forthcoming)
http://oregon.cleanenergymap.com/

Riverside, California: Green Riverside Green Map
www.greenriverside.com/Green-Map-9

Sacramento, California: Solar Sacramento
http://smud.solarmap.org/

Salt Lake City, Utah: Salt Lake City Solar Map
www.slcgovsolar.com/

Santa Clara County, California: Silicon Valley Energy Map
www.svenergymap.org/

San Diego, California: San Diego Solar Map
http://sd.solarmap.org/

San Francisco, California: San Francisco Solar Map
http://sfenergymap.org/

Tallahassee, Florida: Solar Interactive Map
www.talgov.com/you/you-learn-utilities-electric-solar-map.aspx

Solar Mapping Tools

A number of communities have developed online solar mapping tools to educate and inform users about solar technology by estimating the solar energy potential of building sites or open land and providing information about associated benefits. A community can choose to incorporate a wide range of information into its solar map, but three levels of basic input data are needed to begin: local landscape data, meteorological data, and data about financing and incentives (Dean et al. 2009). Additional features may include records on existing systems, links to local installers, photo galleries, news stories, case studies, information on permitting processes and capturing local incentives, schedules of local solar educational offerings, and general information about clean energy technologies.

Communities considering developing a solar map in house should evaluate the qualifications of the staff on hand. Regardless of who is making the map and what technology is being utilized, many stakeholders should be at the table when the map is being developed. Utility providers can provide information on utility rates and who is currently utilizing solar energy. Solar installers can provide information on installation costs and available resource guides. Local government officials can offer information on permitting processes and community goals. Other solar advocates in the community can provide information on local resources, including available solar educational offerings and meetings.

Economic Development Programs

According to research done by the Solar Foundation, BW Research Partnership, and Cornell University, the solar industry is hiring faster than the overall economy and the number of individuals who spend at least 50 percent of their time supporting solar-related activities was up 13 percent in 2012 as compared to 2011; the industry was expected to grow another 17 percent in 2013 (Solar Foundation and BW Research Partnership et al. 2012).

Jobs within the solar sector vary by market segment and include, but are not limited to, installation, manufacturing, technical support, administrative, management, and sales. The majority of solar energy jobs are in the installation, manufacturing, and distribution sectors. These jobs range from highly technical positions, such as engineers and solar system designers, to nontechnical positions in sales, administration, accounting, finance, and law (Solar Foundation and SolarTech et al. 2012). PV products and manufacturing-sector jobs are generally stable, highly skilled, and long term, while other solar jobs can be more variable depending on the long-term stability of the PV market within a community (CH2MHill 2011). Community investments

Figure 7.1. An aerial view of the Aurora Campus for Renewable Energy

SolarTAC

in workforce development programs and business attraction incentives can support the local solar industry in various ways.

Aurora, Colorado, is an example of a local government that has utilized workforce development planning to attract high-paying solar jobs. In 2008 the city created the Aurora Campus for Renewable Energy after it purchased 1,762 acres of land near the Denver International Airport with funds paid out on account of noise violations (Figure 7.1). The campus now hosts the Solar Technology Acceleration Center, the country's largest test facility for solar energy technologies (ICMA 2012).

Programs can target low- and moderate-income individuals and can include job training, certification, and career coaching activities. Solar Richmond in Richmond, California, is a local nonprofit that provides solar industry training, including North American Board of Certified Energy Practitioners certification (Solar Richmond 2013). The program targets low-income individuals and offers career coaching, pilot internships, and free consulting opportunities. Solar Richmond has partnerships with the local government, other nonprofit organizations, and private businesses. Through the work of Solar Richmond, the city government added a 200kW PV installation on its city hall, eliminated permit fees applied to solar installations for Richmond homeowners, and created a solar thermal rebate program.

There are opportunities to partner with local universities, technical institutes, and community colleges to offer coursework and training offerings. For example, the MassGreen Initiative is a program to develop and deliver clean energy workforce training programs at seven different community colleges throughout the state of Massachusetts (Springfield Technical Community College 2013). The program targets a wide range of trainees, including unemployed and underemployed individuals. The program, located at the Springfield Technical Community College in Springfield, Massachusetts, is funded by the Massachusetts Clean Energy Center, in part from funds allocated to the state through the Regional Greenhouse Gas Initiative.

It is important to note that vocational training programs are likely to have the greatest impact in areas with a strong existing solar market. If graduates enter a weak local solar job market, they may be forced to seek employment elsewhere, leading to accusations that local money has been spent with no tangible local benefit.

Educational Programs

There are four major types of training and education programs: pre-employment training, advanced in-service training, continuing education, and "train-the-trainer" activities (CH2MHill 2011). Secondary and post-secondary schools are the primary sources of pre-employment training, while advanced training, continuing education, and train-the-trainer programs may be hosted by a wide variety of institutions and organizations.

Community investments in secondary school curriculums can increase awareness of solar opportunities and benefits as well as lay the groundwork for job-oriented education and training. Local governments can partner with local vocational-technical high schools, colleges, universities, and community-based nonprofit groups to support educational programs that meet the workforce needs of the solar energy sector. Programs can include curriculum and course development, professional development, internship and apprenticeship programs, hands-on instruction training, and dual enrollment programs. In addition, community colleges and local government agencies can create internship programs that facilitate the placement of students and recent graduates who are considering career opportunities in solar energy. Programs can provide paid internships and provide students and companies with the tools to connect and develop a talented pool of young solar energy professionals.

Federal, state, and local grants can support job-training programs directed toward the solar energy industry that move training participants towards financial self-sufficiency. For example, the New York State Energy Research & Development Authority has funded in-state training centers to provide a wide variety of courses on solar PV and other renewable energy and energy efficiency technologies. Students that receive a qualified professional accreditation from these courses are eligible for 50 percent reimbursement for their tuition and examination costs (NYSERDA 2013).

Finally, local governments interested in working with educators to offer solar workforce training can take advantage of train-the-trainer courses provide by the Solar Instructor Training network (SITN) funded by the U.S. Department of Energy. Through the SITN initiative, nine regional trainers assist educational institutions such as community colleges and technical high schools in creating a curriculum for solar system design, installation, sales, and inspection (IREC 2013). As an example, the Rocky Mountain Solar Training Program has assisted Colby Community College in Colby, Kansas, with developing three curriculum options: a 12-hour certificate of completion in solar PV systems, a 36-hour technical certificate for students looking to directly enter the workforce, and an associate's degree program with an emphasis on PV and small wind (Reilley 2013).

SUMMARY

Public investment is the fifth strategic point of intervention for communities looking to promote solar energy use through planning. While most solar development is likely to be initiated by the private sector, public investments provide opportunities for local governments to lead by example. Perhaps the most visible investments that communities can make are public facility construction or retrofitting projects that incorporate solar development. But economic development or educational programs that support local solar market growth provide another way for local governments to expand local solar capacity.

CHAPTER 8

Concluding Thoughts

David Morley, AICP

 As stated in Chapter 1, this report has three primary goals: (1) to provide planners, public officials, and other community stakeholders with a basic rationale for planning for solar energy use; (2) to summarize the fundamental characteristics of the U.S. solar market as they relate to local solar energy use; and (3) to explain how planners, public officials, and other community stakeholders can take advantage of five strategic points of intervention to promote solar energy use.

To reiterate, the five strategic points of intervention are visioning and goal setting, plan making, regulations and incentives, development work, and public investment. Collectively, these points are where planning process participants translate ideas into intentions and intentions into actions.

The preceding chapters covered a wide range of topics and considerations related to planning for solar energy use. While these chapters discuss many specific recommendations and strategies, there are five key themes to guide planners, public officials, and other community stakeholders as they engage in local initiatives to support solar market growth:

• Solar energy is a local resource.

• Local solar markets are sensitive to policy.

• Local plans guide solar energy use.

• Regulatory silence is not the same as support.

• Partnerships can expand local solar opportunities.

SOLAR ENERGY IS A LOCAL RESOURCE

For most communities, solar irradiance is the largest potential local source of energy. Given that a core purpose of local planning is facilitating the development or protection of community resources, it is surprising that few localities acknowledge solar energy as a resource comparable to other local resources such as vegetation, water, minerals, fossil fuel reserves, or historical buildings and cultural heritage sites.

Solar development has environmental benefits. Solar radiation is a carbon-free, emission-free, local fuel source that can help communities meet greenhouse gas reduction goals, energy independence goals, and state or local renewable portfolio standard goals.

Solar development has local economic benefits as well. Installation jobs cannot be outsourced, and solar development helps communities substitute local resources for nonlocal resources. Money spent on locally produced energy stays within the local economy, while money spent on nonlocal energy sources leaves the local economy.

Finally, as with all developable resources, solar development has land-use implications. It requires both space and unimpeded access to solar radiation. As a consequence, solar development affects other types of development as well as the use or conservation of other community resources. Acknowledging solar radiation as a local resource is the first step to successfully balancing competing resource demands.

LOCAL SOLAR MARKETS ARE SENSITIVE TO POLICY

The maturation of solar technology has led to dramatic decreases in installation costs over the past 10 years. With that said, high upfront equipment and installation costs prevent many individuals and business from investing in solar energy systems, and solar electric rates are still higher than conventional fuel rates in most areas of the U.S. As a consequence, it is important to keep in mind the important roles that federal, state, and local policies play in promoting solar market growth.

As with all energy technologies, financial incentives help drive growth in the U.S. solar market. Beyond this, state and local utility regulations and utility business models affect the extent of utility investments in solar energy, when and how PV systems can connect to the grid, and the price an individual utility is willing to pay for the power produced. While all of these factors have received a great deal of attention from solar policy experts, it is only relatively recently that solar advocates have turned their attention to the influence that local land-use and development policy can have on solar development decisions.

While cities and counties typically have little, if any, direct influence on federal and state incentives and utility policies, local governments can still minimize the risk and uncertainty associated with solar development by engendering solar-supportive land-use and development policies. Because most federal and state incentive programs are available for a limited time, an unanticipated development approval delay can prove disastrous for some projects.

LOCAL PLANS GUIDE SOLAR ENERGY USE

Local plans help communities chart courses for more sustainable and livable futures. The goals, objectives, policies, and actions contained in these plans guide local officials as they make decisions that affect the social, economic, and physical growth and change of their communities. For this reason, local plans can play an important role in either promoting or inhibiting solar energy use.

Plans that discuss the local solar resource explicitly and offer clear support for solar development send positive signals to residents and other community stakeholders potentially interested in making solar investments. On the contrary, when plans remain silent about how solar energy use relates to community goals and objectives, the private market may perceive added risk in making solar investments. Furthermore, comprehensive plans, in particular, allow communities to highlight synergies and potential conflicts between solar and other community resources and to summarize any previous, ongoing, and planned policies and actions to support the implementation of goals related to promoting solar energy use.

REGULATORY SILENCE IS NOT THE SAME AS SUPPORT

Surprisingly few communities explicitly sanction solar development through local zoning, subdivision, or building codes. While in some communities this silence has not, traditionally, been viewed as a major barrier to installing accessory solar energy systems, it robs homeowners and other potential installers of certainty about what types of systems are allowed in what locations. Without clear definitions and standards, public officials are forced to make ad hoc use interpretations that can delay or even prevent otherwise routine installations.

Absent specific provisions that enable or encourage solar development, many local codes contain district dimensional and development standards that may unintentionally limit the permissible size or reduce the efficiency of solar systems. These barriers often go undetected until a specific proposal is inadvertently delayed during the approval process. Finally, explicit definitions and standards allow communities to address the potential impacts of solar energy systems on adjacent uses, the natural environment, and community character, thereby avoiding unnecessary controversy that could undermine community support for solar energy use.

An increasing demand for solar gardens and farms translates to increased demand for large sites appropriate for solar energy production. Communities that make a concerted effort to identify and designate appropriate sites for solar farms will likely have a competitive advantage over neighboring localities that take a reactive stance toward large-scale solar development proposals.

Because solar gardens and farms increase opportunities for utility customers to support solar power without purchasing or hosting a system, it is likely that many communities will see a sharp spike in interest in primary-use installations of various sizes. Again, communities that explicitly enable solar gardens and farms through their development regulations may have a leg up on nearby localities that remain silent in their codes.

PARTNERSHIPS CAN EXPAND LOCAL SOLAR OPPORTUNITIES

Local governments may feel they have rather limited opportunities to participate directly in local solar market growth, but development and educational partnerships can expand these opportunities in numerous ways.

While some public entities do elect to own and operate their own solar energy systems, third-party ownership arrangements, where permissible, provide an alternative means for installing solar systems on public facilities.

Through public land disposition and joint development processes, local governments can encourage or require the integration of solar energy systems into private development projects. Furthermore, localities that have mapped the local solar resource and monitor and evaluate land supply can assist private developers in identifying solar development opportunities.

Finally, educational collaborations with local utilities and educational institutions can reinforce a supportive policy framework. Local governments can sponsor or host informational sessions that build awareness about solar incentives and permitting processes, and they can also provide financial assistance, promotional support, or space for workforce development programs that increase local installation expertise.

SUMMARY

Solar energy is a valuable local resource. Before considering specific local policies or actions to promote solar energy use, it is important for planners, local officials, and other community stakeholders to gain a basic understanding of the spatial, technological, economic, and political variables that constrain solar development. Visioning and goal-setting exercises provide the first and best opportunity for residents and other community stakeholders to learn about these variables and to discuss how solar energy use connects to other community goals and values.

When stakeholders identify solar energy as a priority during these exercises, they are influencing the types of plans a community undertakes as well as what will be incorporated into existing plans in the future. Comprehensive, subarea, and functional plans that include solar-supportive goals, objectives, policies, and actions send clear signals to residents, business owners, and other community stakeholders about where and how solar energy use will be sanctioned or encouraged locally. Similarly, communities can provide further certainty to stakeholders by translating plan policies into local development regulations that explicitly enable various types of solar development.

Beyond solar-supportive plans and development regulations, communities can use development and financial incentives, development services, and public-private partnerships to encourage private solar development. And public investments provide opportunities for local governments to lead by example. The common thread running throughout the preceding chapters (and the following appendices) is that planners and planning matter when it comes to capturing the local benefits of solar energy use.

Solar-Friendly Planning System Audit
for Local Governments

Brian Ross

Plan Making Best Practices			
Background information and resource assessment: Plans include identification of community resources and background information that inform the process of defining the desired future outcomes. Recognizing local solar resources as a driver for development in the community helps integrate the resource into decision making. The comprehensive plan is the foundational document, but communities can also address solar development in sub-area and functional plans.			
Solar Best Practice	**Location**	**Yes No**	**Comments**
1. Does the community identify solar radiation as a potentially valuable resource that can drive development in the community?	Plans: Background section, analysis	❏ ❏	
2. Has the community mapped the solar resource or otherwise identified the potential for solar development in the community?	Plans: Background section, analysis	❏ ❏	
3. Has the community identified potential conflicts between solar resources and other resources, such as the urban forest, historic resources, or neighborhood character?	Plans: Background section, analysis	❏ ❏	
Goals and policies: Plans identify the desired future outcomes in the form of goals and policies. Specifically identifying how solar development will benefit the community helps decision makers define how solar resources and solar investment fit with other community resource development or protection goals.			
Solar Best Practice	**Location**	**Yes No**	**Comments**
1. Does the plan identify the economic benefits of solar development?	Plans: Vision, goals, or policies	❏ ❏	
2. Does the plan address climate protection activities or goals?	Plans: Vision, goals, or policies	❏ ❏	
3. Does the plan explicitly support renewable or alternative energy development?	Plans: Vision, goals, or policies	❏ ❏	
4. Does the plan promote the general use or development of local resources?	Plans: Vision, goals, or policies	❏ ❏	
5. Does the plan support the general goal of using built infrastructure (water, sewer, electric and gas utilities, roads) more efficiently?	Plans: Vision, goals, or policies	❏ ❏	
6. Does the community recognize the environmental benefits of solar development (GHG reduction, air quality, minimizing fossil-fuel use)?	Plans: Vision, goals, or policies	❏ ❏	

Development Regulation Best Practices

Solar uses: Solar energy systems can occur as either an accessory use (rooftop or freestanding) or a principal use (solar garden or farm). Identify where solar uses are permitted as-of-right or conditionally. Solar accessory uses are typically permitted as-of-right in any district where buildings are allowed. Principal uses are permitted as-of-right or conditionally in specific appropriate districts.

Solar Best Practice	Location	Yes No	Comments
1. Does the zoning ordinance explicitly permit accessory use solar energy systems?	Zoning: districts, use tables	☐ ☐	
2. Does the zoning ordinance explicitly permit solar energy systems as a principal use in any district?	Zoning: districts, use tables	☐ ☐	

Height limits: Solar resources are sometimes limited to rooftop areas, meaning solar energy systems need to be above the roof (either a flat roof or above the peak of a pitched roof) in order to function. Identify whether solar energy systems are exempted from height limits and the conditions, if any, that the community places on solar energy systems that exceed the height standard. The standard may be different for flat roof buildings than for pitched roof residential buildings.

Solar Best Practice	Location	Yes No	Comments
1. Does the zoning ordinance make height limit exceptions for building systems or equipment such as chimneys, architectural features, rooftop equipment, etc.?	Zoning: General standards or districts	☐ ☐	
2. Does the ordinance identify whether a solar energy system constitutes an exception similar to other building system exceptions?	Zoning: Use-specific standards, general standards, or districts	☐ ☐	

Setbacks/required yards: Solar resources are sometimes limited to areas on the lot that are part of setbacks or required yards in which development is restricted. Identify whether solar energy systems constitute an allowed incursion and the conditions, if any, that the community places on such incursions.

Solar Best Practice	Location	Yes No	Comments
1. Does the ordinance allow incursions (decks, equipment, awnings) into setbacks or required yards?	Zoning: General standards	☐ ☐	
2. Does the ordinance identify conditions or standards under which solar energy systems can extend into setback or required yards?	Zoning: Use-specific standards or general standards	☐ ☐	

Lot coverage: Solar resources may be located in yard areas rather than on rooftops. But accessory freestanding solar energy systems can be restricted due to lot coverage limits, impervious surface limits, or limits on the number of accessory structures per lot. Set standards that allow reasonable solar development in yard areas.

Solar Best Practice	Location	Yes No	Comments
1. Are accessory structures subject to lot coverage limitations or limits on the number of structures per lot?	Zoning: Districts, general standards, or	☐ ☐	

	stormwater standards			
Solar Best Practice	**Location**	**Yes**	**No**	**Comments**
2. Does the ordinance identify whether solar energy systems are subject to coverage standards, and establish conditions under which the community will make exceptions?	Zoning: Use-specific standards, districts, general standards, or stormwater standards	❑	❑	
3. Does the community identify solar collectors as impervious surfaces?	Zoning: Use-specific standards or stormwater standards	❑	❑	

Solar rights: The concept of solar rights includes both a right to install a solar energy system and a right to retain access to direct sunlight over the life of the solar energy system. The combination of solar development zoning standards noted above should be constructed in a manner to clearly identify an as-of-right design for solar energy systems. The as-of-right design will be different for residential and commercial installations, and for rooftop and freestanding installations. Contractors should be able to design a solar installation that is protected by development regulations. System owners should be able to protect long-term access to direct sunlight.

1. Has the community identified characteristics of solar installations that are "as-of-right" for residential and commercial districts?	Zoning: Use-specific standards or solar installation guide	❑	❑	
2. Has the community noted in ordinance that solar development standards are intended to guide and enable development rather than restrict development?	Zoning: Use-specific standards or solar rights standards	❑	❑	
3. Do property owners within common-interest communities have a right to install a solar energy system?	Subdivision or Zoning: Solar rights standards	❑	❑	
4. Do property owners have a means to protect access to direct sunlight over time via a solar easement?	Zoning: Solar rights standards	❑	❑	
5. Does the community, in its development regulation or other ordinances, consider the impact of new development on access to direct sunlight for existing solar systems that are located on adjacent lots?	Zoning: Rezoning or variance standards	❑	❑	

Extra credit: Communities have a number of opportunities to encourage solar development through development regulation.

Solar Best Practice	**Location**	**Yes**	**No**	**Comments**
1. Does the community encourage or require some new buildings to be built "solar-ready"?	Zoning, Subdivision, PUD	❑	❑	
2. Does the community encourage or require in the subdivision process that solar resources be identified or protected in lot configuration, use of	Subdivision or PUD	❑	❑	

	Location	Yes No	Comments
solar easements, building or landscaping standards, or HOA design standards?			
3. Does the community encourage or require development that is part of a public-private partnership (where the community is financial partner via investment, land donation, etc.) to include solar development as part of the project?	Development standards	❑ ❑	

Permitting Best Practices

Transparent permit requirements: Communities should clearly identify the permits needed for solar energy systems in order to create a transparent and consistent development process. All solar energy systems are subject to the state building code and electric code, although not all communities enforce the code or issue permits. Regardless of the local jurisdiction, all PV system contractors must obtain an electric permit prior to commencing work.

Solar Best Practice	Location	Yes No	Comments
1. Does the community issue building permits or enforce the building code?	Building or Community Development dept.	❑ ❑	
2. Does the community or the state electric inspector issue electric permit?	Building or Community Development dept.	❑ ❑	
3. If other permits, such as a separate land-use permit, are required for all solar energy systems, does the contractor use a single permit process for all approvals?	Building or Community Development dept.	❑ ❑	
4. Does the community identify on its website or in print what permits are needed for solar energy systems?	Website or permit counter handouts	❑ ❑	

Building permit process: When communities issue building permits for solar energy systems, the requirements and process for obtaining a permit should be predictable for contractors and counter staff and streamlined when possible (a single process for multiple permits, when multiple permits are necessary).

Solar Best Practice	Location	Yes No	Comments
1. Does the community have written standardized solar building permit application submittal requirements (conditions when a building permit is required, submittal information necessary for obtaining a permit)?	Website or permit counter handouts	❑ ❑	
2. Does the community use the Solar ABCs expedited permit process for small solar energy systems?	Website or permit counter handouts	❑ ❑	
3. For small-scale solar projects, does the community combine application processing into a single process (one	Website, permit counter handouts, or staff	❑ ❑	

	Solar Best Practice	Location	Yes	No	Comments
	permit process for building, structural, zoning)?				
4.	Does the community offer online or over-the-counter submittal and review options for solar energy systems?	Website, permit counter handouts, or staff	❑	❑	

Electric permit process: For communities that issue electric permits for solar development, the requirements for obtaining a permit should be predictable and explicit for both contractors and counter staff. Maximize use of online information specific to solar energy systems, and use online applications if possible.

	Solar Best Practice	Location	Yes	No	Comments
1.	If the community does not issue electric permits, does it track permits or coordinate with state electric inspectors on electric permits issued?	Building or Community Development staff	❑	❑	
2.	Does the community use the Solar ABCs expedited permit process for small solar energy systems?	Building or Community Development staff	❑	❑	
3.	Does the community coordinate with the electric utility on electric and interconnection requirements and inspections?	Building, Community Development, or Utility staff	❑	❑	

Inspections: Coordinate inspections for small solar energy systems.

	Solar Best Practice	Location	Yes	No	Comments
1.	For standard solar development projects does the community require only one inspection for building and electric permits?	Building dept. handouts, website, or staff	❑	❑	
2.	In the inspection process, does the community give contractors a specific time for the inspection (rather than a window of time of an hour or more)?	Building dept. handouts, website, or staff	❑	❑	

Permit fees: Many communities assess permit fees based on the value of the development project. The valuation-based fee is a proxy for estimating the cost of issuing permits and conducting inspections. Solar energy system costs are not, however, indicative of the complexity of the project or the inspection process.

	Solar Best Practice	Location	Yes	No	Comments
1.	Does the community have flat permit fees for any type of building project?	Building dept. handouts, website, or staff	❑	❑	
2.	Does the community have a flat permit fee for small solar energy systems?	Building dept. handouts, website, or staff	❑	❑	
3.	Does the community have a value-based fee structure that excludes the cost of solar collectors, power electronics, or other equipment elements of solar development?	Building dept. handouts, website, or staff	❑	❑	

Solar Energy Goals, Objectives, and Policies in Comprehensive Plans

Darcie White, AICP, *and Paul Anthony*, AICP

City of Andover, Minnesota (pop. 30,598)
Comprehensive Plan, Chapter 1, Foundation of the Comprehensive Plan: Land Use Goals, Objectives, and Policies (2008)
http://files.andovermn.net/pdfs/Planning/CompUpdate/2008%20Comp%20Plan_
Final%20Approved%20Documents/Chapter%20One_Foundation%20of%20the%20
Comprehensive%20Plan.pdf

Goal: Protect and develop access for alternative energy systems
Objective: Preserve reasonable access to all parcels so that alternative forms of energy can be used to supplement or replace conventional forms of energy
Policies:

- Encourage and support educational programs and research that focuses on alternative or renewable energy systems such as offered by Metro Cities, University of Minnesota Extension Services, Minnesota Office of Environmental Assistance, Anoka County and other organizations
- Encourage the possible use of solar energy in future housing developments
- Encourage future site and building plans to design for efficient use of solar energy including such elements as the location of windows, shade trees, windows, and driveways

City of Brandenton, Florida (pop. 49,546)
Comprehensive Plan, Coastal Management and Conservation Element
http://bradenton.govoffice.com/vertical/Sites/%7B2D1C3C91-86C5-4ACC-86B6-
6CFA76381D46%7D/uploads/%7B73F2C88D-E782-4159-B6DD-F41BB22D373E%7D.PDF

Goal 7: Energy Efficiency and Conservation
Objective 7.4: Solar Energy
    The City will promote, support and require, as appropriate, the use of solar energy.
Policy 7.4.1: Solar Ready Buildings
    The City will require where feasible, all new buildings be constructed to allow for easy, cost effective installation of solar energy systems in the future, using such "solar-ready" features as:
- Designing the building to include optimal roof orientation with sufficient south-sloped roof surface,
- Clear access without obstructions (e.g., chimneys, heating and plumbing vents) on the south sloped roof;
- Designing roof framing to support the addition of solar panels;
- Installation of electrical conduit to accept solar electric system wiring; and
- Installation of plumbing to support a solar hot water system and provision of space for a solar hot water storage tank.
Policy 7.4.2: Passive Solar Design
    The City will require that any building constructed in whole or part with City funds incorporate passive solar design features.
Policy 7.4.3: Protection of Solar Elements
    The City will protect existing, active and passive solar design elements and systems from shading by proposed neighboring structures and landscape elements.

City of Fort Collins, Colorado (pop. 143,986)
City Plan Fort Collins, Energy (2011)
www.fcgov.com/planfortcollins/pdf/cityplan.pdf

Principle ENV5: To reduce net community energy use for new construction from conventional fossil fuel sources, the City will expand on current efforts and develop new strategies for increased energy efficiency and use of renewable energy.

Policy ENV 5.2 – Utilize Solar Access. Protect unobstructed sunlight in planning and development processes to promote the use of solar energy.

Policy ENV 5.4 – Support Renewable Energy in New Development. Support the use of renewable energy resources in the layout and construction of new development.

Policy ENV 5.6 – Update Regulations. Regularly update codes that define minimum acceptable community standards for new construction with regard to energy efficiency and renewable energy use.

Policy ENV 5.7 – Offer Incentives. Offer a variety of monetary and other incentives to encourage new construction to substantially exceed minimum code requirements for energy efficiency and renewable energy use.

Policy ENV 5.8 – Participate in Research, Development and Demonstrations. Participate in research, development and demonstration efforts to remain at the forefront of emerging technologies and innovative solutions regarding the energy performance of new construction.

Jackson County, Oregon (pop. 203,206)
Comprehensive Plan, Chapter 11, Energy (2007)
www.co.jackson.or.us/Files/11%20-%20ENERGY.pdf
POLICY: ENERGY CONSERVATION MEASURES SHALL BE UTILIZED IN NEW DEVELOPMENT PROJECTS TO ACHIEVE ENERGY EFFICIENT DEVELOPMENT THROUGH COMBINATIONS OF SITE PLANNING, LANDSCAPING, BUILDING DESIGN AND CONSTRUCTION PRACTICES.

C) Establish optional thermal efficiency performance standards for structures designed to utilize passive solar space heating techniques based upon the concepts embodied in the City of Davis, California's code and modified to reflect the southern Oregon climate.

D) Devise and amend applicable codes and ordinances to foster the alignment of streets which maximize opportunities for solar orientation of structures.

E) Revise and amend applicable codes and ordinances to provide flexible setback requirements conducive to the solar orientation of structures. Develop a solar easement ordinance to guarantee access to incident solar radiation for property owners, except where preexisting conditions preclude such access. Revise and amend applicable codes and ordinances to assure the integration of solar access protection provisions as provided for in Senate Bill 299 (1979 Oregon Laws Chapter 671).

I) Prepare and distribute a developer's energy conservation handbook and guide manual delineating energy conserving principles and techniques covering site design, climatic factors of southern Oregon, solar orientation, landscaping for wind and sun control, building design and construction, alternative energy systems and devices, and other energy conserving/efficiency factors.

J) Revise applicable regulatory requirements in codes and ordinances to permit or require where appropriate, the use of on-site renewable energy facilities, including individual homesite and district heating and cooling, and integrated community/neighborhood renewable energy generation systems whose energy sources include solar, hydro, wind, biomass, and geothermal.

POLICY: THE COUNTY SHOULD BE MORE ENERGY SELF-SUFFICIENT AND SHALL ACTIVELY ENCOURAGE THE DEVELOPMENT AND USE OF LOCAL RENEWABLE ENERGY RESOURCES AND ALTERNATIVE ENERGY SYSTEMS ON THE COMMUNITY, NEIGHBORHOOD, AND INDIVIDUAL HOMESITE LEVEL.

A. Adopt the following as an interim set of performance standards for solar space and water heating and cooling systems until such time as more definitive and applicable standards become available:

> i) Solar Heating and Domestic Hot Water Systems, HUD Intermediate Minimum Property Standards Supplement, U.S. Department of Transportation and Urban Development.
>
> ii) Interim Performance Criteria for Solar Heating and Cooling Systems in Commercial Buildings, National Bureau of Standards (Center for Building Technology and Institute for Applied Techniques) prepared for Energy Resource and Development Administration, Division of Solar Energy, Washington, D.C.

C) Encourage private lending institutions to give incentives for the utilization of alternative renewable energy resources and systems.

D) Require the use of solar energy to heat swimming pools, except in cases of therapeutic necessity.

F) Revise and amend the zoning ordinance to incorporate alternative energy systems and devices.

G) Institute a long-term continuous and action-oriented energy planning program which places a high priority on citizen involvement.

H) Investigate the feasibility of utilizing the public utility districts of local improvement or special district concept as a means of encouraging energy conservation and self-sufficiency. This could be applicable for developing a solar utility district.

I) Encourage destination resorts to make use of on-site renewable energy resources.

Town of Jackson, Wyoming (pop. 9,577)

Jackson/Teton County Comprehensive Plan, Common Value 1, Ecosystem Stewardship (2011)
www.tetonwyo.org/compplan/docs/2011/07/120406_JacksonTeton_Part2_CV-1_
EcosystemStewardship.pdf

Principle 2.1: Reduce consumption of non-renewable energy
Policy 2.1.c: Increase local use and generation of renewable energy

> Using solar, wind, geothermal, and/or hydro energy that has less impact to the climate is the community's preference. The community will work with local utilities and other agencies, nonprofits, and businesses to identify local renewable energy generation opportunities so that it is not necessary to add non-renewable energy sources to the community's energy portfolio. Integration of renewable energy into the community's energy portfolio should be done consistently with the community's Vision.

Policy 2.1.d: Allow and encourage onsite renewable energy generation

> Production of energy from renewable sources on individual properties should be allowed and encouraged. The transmission of electricity is extremely inefficient. Reducing that component of our energy infrastructure could result in a large cumulative decrease in demand for non-renewable energy. Exemptions to Town and County regulations should be considered to facilitate the installation of on-site renewable energy sources. The community will also explore incentives for on-site renewable energy, utilizing best available practices.

City of Lancaster, California (pop. 156,663)

General Plan, Section II, Plan for the Natural Environment (2009)
http://www.cityoflancasterca.org/Modules/ShowDocument.aspx?documentid=9323

Policy 3.6.2: Encourage innovative building, site design, and orientation techniques which minimize energy use.

> Specific Action 3.6.2(b): Review and revise as necessary development code provisions for the application of energy conservations measures in subdivisions, site plans, conditional use permits and other land use entitlements. The provisions could include solar access through lot size, configuration and orientation; building height, setbacks, and coverages; renewable energy resource systems with permitted and accessory uses; and other innovative measures promoting energy efficiency.

Policy 3.6.3: Encourage the incorporation of energy conservation measures in existing and new structures.

Specific Action 3.6.3(a): Investigate the feasibility of adopting an Energy Ordinance that will encourage the installation of energy conservation measures on rehabilitation or expansion projects; and retrofitting energy conservation measures on existing structures that require major renovation. Specific measures include, but are not limited to, solar heating systems for pools and other appropriate facilities and provisions for industrial projects that will facilitate the installation of photovoltaic electric generating units.

Specific Action 3.6.3(b): Explore the feasibility of requiring solar systems in new residential and non-residential construction. If practical, amend the municipal code to address requirements for solar energy use.

Specific Action 3.6.4(c): Promote the application of active solar energy systems by facilitating the efforts of Federal and State entities in the allocation of cost incentive programs.

City of Livermore, California (pop. 80,968)
General Plan, Climate Change Element (2009)
www.cityoflivermore.net/civicax/filebank/documents/6103/

Climate BMP No. 3 – Incorporate solar roofs into commercial development. Residential development to be "solar-ready" including proper solar orientation (south facing roof area sloped at 20° to 55° from the horizontal), clear access on the south sloped roof (no chimneys, heating vents, plumbing vents, etc.), electrical conduit installed for solar electric system wiring, plumbing installed for solar hot water system, and space provided for a solar hot water storage tank.

Energy Policy P.1 - ALTERNATIVE ENERGY DEVELOPMENT PLAN: Explore possibilities for alternative energy production and establishing City-wide measurable goals. Develop an Alternative Energy Development Plan to identify the allowable and appropriate alternative energy facility types (i.e., solar photovoltaic (PV) on urban residential and commercial roofs and wind farms on the edge of town or in rural areas) and locations within Livermore as well as propose phasing and timing of alternative energy facility and infrastructure development. Continue to identify and remove regulatory or procedural barriers to producing renewable energy in building and development codes, design guidelines, and zoning ordinances. . . .The Alternative Energy Development Plan shall identify optimal locations and best means to avoid noise, aesthetic, and other potential land use compatibility conflicts (e.g., installing tracking solar PV or angling fixed solar PV in a manner that reduces glare to surrounding land uses.)

City of Morgan Hill, California (pop. 37,882)
Morgan Hill General Plan, Open Space and Conservation (2010)
http://www.morgan-hill.ca.gov/DocumentCenter/Home/View/1148

Goal 7. Conservation of natural resources policies

7i. Use of renewable energy generation opportunities should be evaluated for all existing and future public buildings and facilities.

7j. The incorporation of renewable energy generating features, like solar panels, should be encouraged in the design of new development and in existing development.

Actions

7.3 Develop local ordinances that promote energy conservation and efficiency. Examples of such ordinance include: energy audits, solar access, solar swimming pool heating, insulation and solar retrofit, and solar water heating.

7.4 Establish programs under HCD Block Grant rehabilitation or Section 220 funds, to weatherize and solar retrofit existing homes.

7.6 In compliance with Section 66473.1 of the State Subdivision Map Act, promote subdivision design that provides for passive solar heating and natural cooling through the Development Review Committee subdivision review procedures.

Pinal County, Arizona (pop. 375,770)

We Create Our Future: Pinal County Comprehensive Plan, Chapter 7, Environmental Stewardship (2012)

http://pinalcountyaz.gov/Departments/PlanningDevelopment/Comprehensive PlanUpdate/Documents/00.Comprehensive%20Plan%202012.pdf

Conservation

Policies:

7.3.1.3 Locate solar energy generation equipment on County facilities which cost/benefit analyses proves advantageous.

7.4.2.1 Encourage developments that use energy smart site design (e.g., solar orientation, cluster development).

Renewable Sources

7.6 Goal: Expand renewable energy in Pinal County.

7.6.1 Objective: Support small scale renewable energy projects

Policies:

7.6.1.1 Support statewide policy that provides property tax credits for renewable energy facilities on individual homes and businesses from net assessed valuation calculations

7.6.1.2 Assess current codes so they are supportive in permitting small scale renewable energy projects. Explore ways to reduce barriers caused by homeowner's association restrictions.

7.6.1.3 Work with developers and energy providers to design neighborhoods with optimum solar orientation.

7.6.1.4 Support state and federal incentive programs for the development of renewable energy infrastructure for individuals and businesses.

7.6.1.5 Develop/amend ordinances to protect solar access through sensitive building orientation and for property owners, builders and developers wishing to install solar energy systems.

7.6.1.6 Support the transmission of renewable energy from sources within and outside of Pinal County.

7.6.2 Objective: Support the growth of the renewable energy in Pinal County.

Policies:

7.6.2.1 Identify through specific area planning potential locations for renewable energy projects.

7.6.2.2 Support the attraction of renewable energy providers through the County's economic development strategy.

7.6.2.3 Work with economic development proponents to develop education and training programs for renewable energy employment opportunities.

City of Pleasanton, California (pop. 70,285)

Pleasanton General Plan 2005–2025, Energy Element (2009)

www.ci.pleasanton.ca.us/pdf/genplan-090721-energy.pdf

Policy 7: Promote renewable energy.

Program 7.1: Encourage public and private entities to generate renewable energy.

Program 7.2: Use solar in public facilities and encourage the use of solar in private facilities, where feasible and cost effective.

Program 7.3: Promote and encourage photovoltaic demonstration projects in association with public or private development.

Program 7.4: Study the feasibility of starting or joining a photovoltaic co-op program and explore related financial considerations.

Program 7.5: For new construction, require roofs that are strong enough and have roof truss spacing to hold photovoltaic panels, where feasible and cost effective.

Program 7.6: Require solar water heating and/or photovoltaic-ready roofs in new construction, i.e., roofs with wiring installed for a roof-mounted photovoltaic system, where feasible.

Program 7.7: Support the production of alternative and renewable fuels and fuelling stations in Pleasanton.

Program 7.8: Consider a photovoltaic joint venture project on private property.

Program 7.9: Work with the City of Livermore and Spectrum Energy to develop a solar cities program or standardized solar-energy-system-installation designs for residences and potentially for businesses.

Program 7.10 : Explore the concept of funding energy efficiency upgrades for residential and commercial buildings as authorized by AB 811.

Town of Sahuarita, Arizona (pop. 25,259)
General Plan (2003)
http://sahuaritaaz.gov/DocumentCenter/View/101

Goal LU-1: Promote an orderly, directed, and balanced land use pattern that recognizes the rural character of the community, while meeting the housing, services, employment, and recreational needs of the Town of Sahuarita.
Objective:
LU-1.2: Promote land use patterns that conserve natural resources including land, open space, air quality, water quality and quantity and energy.
Policies:
LU-1.2.7: Promote the orientation of new housing stock to maximize use of solar energy and review building codes to ensure that new structures utilize best available practices for energy conservation.

Goal PFS-1: Provide a high level of public facilities, utilities and services to support and efficiently serve the Town.
Objective:
PFS-1.5: Promote coordination among agencies for maximum efficiency in the delivery of public services to the Town.
Policies:
PFS-1.5.5: Encourage utility providers to consider the use of solar power and other renewable resources.

Goal ENV-1: Manage the natural resources of the area in a manner that will balance their ecological value and economical, aesthetic and safety potential.
Objective:
ENV-1.6: Promote energy efficiency through conservation and the use of alternative energy practices.
Policies:
ENV-1.6.1: Encourage the use of solar energy or other appropriate energy conservation technologies, rainwater harvesting and other renewable resource practices.

Solar Strategies, Measures, and Actions in Functional Plans

City of Albany, California (pop. 18,539)
Climate Action Plan (2010)
www.albanyca.org/Modules/ShowDocument.aspx?documentid=11490

Measure BE 1.1: Install cost-effective renewable energy systems on all City buildings, and install building performance data displays to demonstrate savings.
Measure BE 2.1: Develop comprehensive outreach programs to encourage energy efficiency and renewable energy investments in the community.
Measure BE 2.2: Identify and develop low-cost financing products and programs to encourage investment in energy efficiency and renewable energy within existing residential units and commercial buildings.
Measure BE 2.3: Develop and implement point-of-sale residential and commercial energy efficiency upgrade requirements.
Measure BE 2.4: Identify and facilitate solar energy EmPowerment districts in commercial, industrial and mixed-use portions of the city.

Town of Arlington, Massachusetts (pop. 42,844)
Arlington Sustainability Action Plan, Volume I: Climate Action Plan, Section 3, Energy Sourcing
http://ase.tufts.edu/uep/degrees/field_project_reports/2005/8-arlington_sustainablility_action_plan.pdf

Proposed Municipal Measures:
3.3.3.2 Installation of PV Systems on Municipal Buildings:
  The Town should make use of the available grants for the installation of photovoltaic (PV) systems on municipal buildings. Specifically, the Town should expand the solar project underway at Arlington High School to provide a larger portion of the buildings electricity, conduct associated classroom activities to raise awareness and engage students in the process. The Town should also consider similar systems for other schools.
3.3.4.1 Installation of Residential PV Systems:
  Residents can take advantage of some of the funds and technical assistance offered by the Small Renewables Initiative. In addition to the MTC's facilitation, there is a program in Massachusetts, called net metering, for those installing renewable energy systems smaller than 60 kW in size (the typical home uses 4-6 kW; local businesses may range from this to beyond 100 kW).
3.3.4.2 Installation of PV Systems in Commercial Buildings:
  Business owners can take advantage of the funds and technical assistance offered by the Small Renewables Initiative and the Commercial, Industrial & Institutional Initiative through the MTC. The Town of Arlington should create and publicize an "Arlington Alliance of Sustainable Businesses" program that encourages local businesses, such as supermarkets, auto dealerships, and other commercial buildings, to take advantage of the rebates and tax deductions that are available for residents who install renewable technologies.
3.3.4.4 Residential Use of Solar Hot Water Heating:
  Water heating accounts for approximately 14 percent of the average family's home energy consumption in the USA. Solar hot water heating systems can help residents cut water heating energy use by 40 to 60 percent. A solar hot water heating system collects thermal energy from the sun to heat the water used to take showers, wash dishes

and clean laundry. The Town can start initiatives for the installation of residential use of solar hot water heating systems.

City of Hayward, California (pop. 114,186)
Climate Action Plan (2009)
www.hayward-ca.gov/GREEN-HAYWARD/CLIMATE-ACTION-PLAN/pdfs/2009/
CAP_Final/Hayward_CAP_FINAL_11-6-09%20-%20full%20document.pdf

Strategy 5 – Energy: Use Renewable Energy
Action 5.1: Develop a program for the financing and installation of renewable energy systems on residential buildings including single and multiple family residential buildings and mobile homes.
Action 5.2: Develop a program for the financing and installation of renewable energy systems on commercial buildings.
Action 5.3: Incorporate a renewable energy requirement into Private Development Green Building Ordinance and the Residential and Commercial Energy Conservation Ordinances.
Action 5.4: Increase the renewable portion of utility electricity generation by advocating for increased state-wide renewable portfolio standards; and consider participating in community choice aggregation, or other means.
Action 5.5: Conduct a city-wide renewable energy assessment to estimate the total renewable energy potential and costs and benefits of developing that potential within City bounds. Develop a plan for capturing all cost-effective opportunities.
Action 5.6: Ensure that all new City owned facilities are built with renewable energy (i.e. PV and/or solar hot water) systems as appropriate to their functions.

City of Minneapolis, Minnesota (pop. 382,578)
Minneapolis Climate Action Plan: Steering Committee Recommendation, Buildings and Energy (2013)
www.minneapolismn.gov/www/groups/public/@citycoordinator/documents/webcontent/wcms1p-109331.pdf

Renewable Energy
1. Support efforts to align utility practices with City and State renewable energy policy.
2. Implement small to mid-sized business renewable and on-site renewable incentive programs.
3. Investigate the feasibility of large-scale renewable energy purchasing for municipal government and/or residents.
   • Create policies and programs to promote readiness for renewable energy into all new commercial and residential buildings.
   • Develop a "solar-ready" building certification.
4. Encourage "net-zero" energy buildings.
5. Support new financing and ownership models for developing Minneapolis' solar resource.

City of New Rochelle, New York (pop. 77,062)
greeNR: The New Rochelle Sustainability Plan 2010-2030, Part 1, Energy and Climate
www.newrochelleny.com/DocumentCenter/Home/View/2054

1.5 Renewable Energy Generation
SHORT-TERM RECOMMENDATIONS - COMPLETE BY YEAR 3
   (1) Examine renewable energy technologies to determine which are appropriate for private installation and use in New Rochelle. Ensure input from experts and neighborhood association representatives. Also define dimensional screening restrictions necessary to limit visual or noise impacts.
   (2) Amend the New Rochelle building and zoning codes to accommodate the forms of renewable energy production deemed appropriate.

MEDIUM-TERM RECOMMENDATIONS – COMPLETE BY YEAR 10

(1) Conduct an inventory of public buildings and public land to identify locations that may be suitable for renewable energy generation. Conduct feasibility and financial analyses to determine the costs and benefits of City-funded renewable energy projects. Also explore options for leasing or licensing public property to private energy producers, including solar power purchase agreements. Adopt and begin to implement a renewable energy generation plan based on these analyses. Reach out to the School District to gauge interest in a similar analysis of School buildings and properties.

(2) Consider the creation of an electric CHP (Combined Heat & Power), solar-powered or wind-powered charging station at the New Rochelle Transit Center to facilitate the use of electric vehicles by commuters and other drivers. If feasible, then implement as local resources and/or the availability of grants permit.

(3) Continue reviewing local Building and Zoning Codes to determine whether new amendments are required to address evolving renewable energy technology.

(4) Advocate for the purchase of renewable energy by utilities and State authorities.

LONG-TERM RECOMMENDATIONS – COMPLETE BY YEAR 20

(1) Continue reviewing local Building and Zoning Codes to determine whether new amendments are required to address evolving renewable energy technology.

(2) Continue to implement plans for renewable energy production on public land and in public buildings.

City of Novato, California (pop. 51,904)

Novato Climate Change Action Plan (2009)

www.ci.novato.ca.us/Index.aspx?page=1386

GOAL 2: RENEWABLE ENERGY

Reduce emissions associated with energy generation through promotion and support of renewable energy generation and use.

Measure 6: Municipal Renewable Energy: Install cost-effective renewable energy systems on all buildings and facilities and purchase remaining electricity from renewable sources.

Measure 7: Community Renewable Energy Facilitation: Identify and remove barriers to small-scale, distributed renewable energy production within the community.

- Adoption of incentives, such as permit streamlining and fee waivers, as feasible.
- Amendments to development codes, design guidelines, and zoning ordinances, as necessary.
- Creation of an "AB 811" or municipal financing program for small and large projects.

Orange County, Florida (pop. 1,145,956)

Climate Change Plan for Orange County Government (2007)

www.broward.org/NaturalResources/ClimateChange/Documents/Orange%20 County%20FL%20Climate%20Sustainability%20Plan.pdf

Goal 1: Adopt policies to establish and implement a County Renewable Energy Initiative

Objectives:

A. Apply for grants and commit funds for solar photovoltaic (PV) panels at the Orange County Convention Center (OCCC). This, up to 1-megawatt, system could be the largest array of solar photovoltaic panels in the southeast. This will help the OCCC reduce their energy consumption from fossil fuel.

B. Develop a program to provide tax incentives and/or tax credits for solar energy manufacturers within the county.

C. Retrofit county buildings with renewable energy systems. This supports hurricane mitigation efforts to have decentralized energy available. Evaluate a specific goal of having 15% of power from all county owned buildings from alternative energy sources within 15 years.

D. All new county buildings meet the Leadership in Energy and Environmental Design (LEED) standards (originally established by Mayor Crotty in September 2005 commitments).

E. Partner with electric utilities to develop green power programs. Sell renewable energy credits (RECs) from the OCCC project to generate more alternative energy on county owned buildings.

F. Consider establishing an incentive program to increase solar hot water heater and PV panels on residential homes and businesses within the county.

City of Tulare, California (pop. 59,278)
City of Tulare Climate Action Plan (2011)
www.ci.tulare.ca.us/pdfs/departments/planning/City_of_Tulare_CAP_2011.04.11_complete.pdf

Goal 2. Promote and support renewable energy generation and use.
MEASURE RE 2.2: RENEWABLE ENERGY FOR COMMERCIAL AND INDUSTRIAL FACILITIES

Increase reliance on local renewable energy sources through provision of a minimum of 30% of commercial and industrial energy needs from on-site renewable energy sources by 2030.

ACTIONS FOR MEASURE RE 2.2:

RE 2.2.1 Develop a renewable energy strategy that encourages installation of solar energy systems through streamlined permit procedures, optional CALGreen Tier 1 measures, adoption of incentives, fee waivers, or a municipal finance district program that provides a low-risk option for property owners to invest in on-site renewable energy installations.

RE 2.2.2 Continue to participate in the second phase of the statewide AB 811 program, the California PACE Program, to achieve the provision of renewable energy.

RE 2.2.3 Encourage participation in Energy Star programs and best practices for commercial and industrial buildings.

RE 2.2.4 By 2019, require new commercial and industrial land uses greater than 5,000 square feet in size to utilize on-site renewable energy systems to offset a minimum of 30% of the projected building energy use or to pay an in-lieu fee or similar offset fund to be established by the City. Renewable energy systems may include energy generated by solar, wind, geothermal, water, or bio-based energy capture systems.

RE 2.2.5 Encourage private development of a community solar group buy program.

MEASURE RE 2.4: RENEWABLE ENERGY FOR RESIDENTS

Increase reliance on local renewable energy sources through provision of a minimum of 15% of baseline residential energy needs from on-site renewable energy sources by 2030.

ACTIONS FOR MEASURE RE 2.4:

RE 2.4.1 Implement the Tulare Affordable Solar Program (TASP).

RE 2.4.2 Investigate additional funding sources for the TASP to provide funding mechanisms targeted to the City's affordable housing stock.

RE 2.4.3 Identify barriers to use of on-site renewable energy for residential uses.

RE 2.4.4 Develop a renewable energy strategy that encourages installation of solar energy systems through streamlined permit procedures, optional CALGreen Tier 1 measures, adoption of incentives, fee waivers, or a municipal finance district program that provides a lowrisk option for property owners to invest in on-site renewable energy installations. (See also RE 2.2.1.)

RE 2.4.5 Continue to participate in the second phase of the statewide AB 811 program, the California PACE Program.

RE 2.4.6 Identify partners and encourage private sector initiatives to sponsor residential community solar projects or solar group buy efforts.

Model Solar Development
Regulation Framework

Darcie White, AICP, and Paul Anthony, AICP

This appendix is intended to serve as a tool to assist communities in translating existing plans and policies into clear and enforceable regulations that support solar energy use. It establishes a framework to guide the development of solar regulations that meet the unique needs of each community and are fully integrated within a community's existing regulations. This integrated approach is intended to create a more level playing field for solar energy systems as a distinct use as well as to avoid potential conflicts with other regulations. The framework may also be used to develop a stand-alone solar development ordinance that addresses the same topics; however, it is essential that communities carefully review any new provisions recommended as part of a stand-alone ordinance against existing regulations to avoid conflicts.

Solar energy systems come in a wide variety of shapes and sizes and may be installed in a multitude of development contexts—ranging from multi-acre solar installations surrounded by undeveloped public land to small-scale systems located in established residential neighborhoods. Each type of system and each distinct development context brings with it a different set of site planning considerations and potential impacts. As a result, solar development regulations must be carefully tailored to address a broad range of local considerations. Communities adopt solar development regulations for three basic reasons:

- *To reinforce community support for solar energy use*—Adopting regulations that clearly define where solar energy systems are allowed, where they are not allowed, and what regulations apply increases predictability for residents and businesses who wish to install solar energy systems. It also increases predictability for the solar installers who typically navigate the review and approval process.
- *To identify and remove potential barriers to solar energy use*—A community's development regulations may contain a variety of direct or indirect barriers to solar energy use. Indirect barriers are issues that a code is silent on and may consequently make a solar application difficult to submit or process, such as a lack of definitions for different types of solar energy systems. Direct barriers specifically limit solar energy systems in terms of their size or location, such as through the prohibition of solar energy systems in one or more zoning districts or through specific development standards, such as setbacks or lot coverage restrictions. Often, these barriers go undetected until a solar application is inadvertently delayed during the approval process.
- *To minimize potential impacts of solar energy systems*—Regardless of how aggressively a community wishes to encourage solar energy use, its development regulations must still address the potential impacts of solar energy systems on adjacent uses, the natural environment, and community character.

STEP 1: REVIEW AND ASSESS EXISTING REGULATIONS

The first step for any community in formulating solar development regulations is to review and assess what rules are on the books today at the state and local level and to identify a list of issues to be addressed as part of the process. This process should result in a brief summary report that can serve not only as the foundation for the development of new regulations, but as a foundation for discussion with key stakeholders about the community's intent and line of thinking with regards to each proposed regulation. Taking the time to conduct necessary research, identify key issues, and establish a framework for the new regulations upfront will ensure discussions with residents, solar installers, business owners, elected and appointed officials, and other stakeholders are well informed and productive and will ultimately lead to a more streamlined review and adoption process.

Start by reviewing applicable state statutes to determine whether any specific authorizations or prohibitions specific to solar energy use currently exist. For example, statutes in some states, such as North Carolina, forbid local government regulations from prohibiting or effectively prohibiting the installation of solar collectors. Others, including those in Colorado, California, and Arizona, preempt private covenants that prohibit solar energy systems. These and other potential limitations should be taken into account, as applicable, to ensure compliance at a local level.

Next, review and analyze existing codes and regulations to determine if and how solar is addressed today. At a minimum, codes and regulations should achieve the following objectives:

- Clarify which types of solar energy systems are allowed and where
- Mitigate potential compatibility issues and nuisances associated with solar equipment, such as height allowances, visual impacts, and encroachment
- Define and protect solar access

The following sections provide an overview of specific development regulations that should be examined as part of the review and assessment process to address the objectives outlined above. These code provisions include permitted uses, dimensional standards, development standards, and definitions (See Table D.1.).

***Table D.1.** Baseline considerations for developing solar regulations*

Source: Work for hire by authors for this report

Topic to be Addressed	Key Considerations
Permitted Uses	Types of solar energy systems permitted as a primary vs. accessory use; zoning districts in which different types of solar energy systems are permitted
Dimensional Standards	Height, lot coverage, and setbacks applicable to solar energy systems
Development Standards	Screening, placement (on building or side), and site planning for solar access (lot and building orientation)
Definitions	Types of solar energy systems, solar access considerations, and related terminology

Permitted Uses

With the exception of some form-based codes, most zoning codes include a section on permitted or allowed uses. This section of the code generally defines the types of uses that are allowed in different zoning districts as either primary, conditional, or accessory uses. Some codes also include use-specific dimensional or development standards.

Typical Regulatory Issues. Although building-mounted solar energy systems are sometimes addressed as potential building appurtenances, along with antennae, satellite dishes, chimneys, and other common features, most codes do not explicitly define solar energy systems as a primary or accessory use. This silence within the code creates uncertainty for applicants, planners, and residents as to the types of solar energy systems that would be permitted in different locations, what regulations would be used to review potential applications, and how long the review process might take.

Key Questions.

- Are solar energy systems currently allowed by right as a primary or accessory use?
- If so, what types of solar energy systems are allowed and in which zoning districts?
- Are these allowances consistent with adopted policies and with solar ordinance objectives? If not, what types of amendments are necessary?

Special Considerations. Absent clear policy guidance that specifically limits solar energy systems to one or more particular areas of the community, many communities allow accessory systems by right in most, if not all, zoning districts (often subject to specific dimensional or development standards). This is particularly true in communities with significant solar resources and strong political support for renewable energy. However, outside of these communities there is considerable variation in use permissions. These variations are typically based on the type and size of solar energy system and whether the system functions as a primary or accessory use. In general, solar energy systems that function as a primary use—commonly referred to utility-scale solar energy systems, solar gardens, or solar farms—are less likely to be permitted by right in all districts.

Primary-use installations are typically ground- or pole-mounted and range from less than one acre in size to as large as 40 acres or more. Although the scale of the equipment used for such installations also varies, most include equipment that is taller than that typically used in a residential or commercial context and significantly larger solar collectors. Potential impacts associated with the larger scale of such solar installations and equipment can also be more pronounced. Water quality, glare, and even noise generated from the motors used to align the collectors for maximum efficiency are typical issues to consider. As a result, some communities choose to limit this type of system to nonresidential districts or more rural locations where the potential for impacts on adjacent uses is lower. Other communities, such as Milwaukee, have fully embraced solar as an integral part of an overall renewable energy strategy and allow solar gardens or farms by-right within a range of residential, commercial, and industrial zoning districts.

If clear policy direction does not exist to inform this issue locally, take into account the location, size, number, and surrounding context of sites within the community that are potentially suitable for different types of solar installations and identify possible impacts associated with each as part of the review and assessment process. Use this analysis to identify potential conflicts and determine what types of solar energy systems should be allowed in each of your community's zoning districts.

Dimensional Standards

Dimensional standards establish the basic parameters that uses must adhere to within specific zoning districts. These parameters commonly include maximum height, maximum lot coverage, and minimum setbacks from property lines.

Typical Regulatory Issues. There are two primary regulatory issues related to dimensional standards for solar energy systems. First, a lack of dimensional standards may increase the potential for poorly sited solar energy systems. Second, dimensional standards may exist, but may not be effective or may apply in unintended ways. In some cases, existing dimensional standards are too restrictive and serve as a barrier to solar, or alternatively, dimensional standards may be too flexible, resulting in compatibility issues in some locations.

Key Questions.

- Does the code contain dimensional standards that guide different types of solar installations (e.g., minimum setbacks from property lines or adjacent residential uses, limits on the height of roof appurtenances, or maximum equipment height)?

- If so, have specific issues been raised in the administration of these standards? For example, have applications for roof-mounted solar energy systems ever been denied because they exceed maximum height limits by a modest amount? Or, have solar gardens or farms experienced challenges in the review process that would have been alleviated by the presence of clear, enforceable standards?

- If not, what types of dimensional standards should be included to address different types of solar installations? How should these standards vary by zone district, if at all? Should solar energy systems be allowed greater flexibility in dimensional standards (e.g., reduced setbacks or increased lot coverage) in deference to their alignment with economic development or sustainability goals?

Special Considerations. Appropriate dimensional standards may vary greatly depending upon the type and size of solar energy system and the location in which it is installed. As a general rule, the more sensitive the surrounding development context the more restrictive dimensional standards typically are. The following contexts may merit more restrictive standards:

- *Residential zones or historic districts*—More restrictive standards may be appropriate in some residential zones or historic districts where community character is of particular concern. In this type of setting, some communities limit the placement of solar energy systems to roof planes or rear yard locations that are not visible to the street. Others take a more flexible approach, stating simply that these locations are "preferred" without prohibiting them outright. The key is to reconcile any such limitations with state statutes to ensure local regulations do not conflict.
- *Environmentally sensitive or rural zones*—An environmentally sensitive site in a rural location may warrant broader setbacks or other standards to protect specific natural resources, such as a stream or documented wildlife corridor. In zoning districts with environmentally sensitive sites that vary greatly in terms of size, condition, and resources to be protected, it may make sense to require a special use permit for solar energy systems in order to verify appropriate site conditions prior to approval.

While the examples outlined above focus on the potential need for more restrictive dimensional standards, other communities who wish to encourage solar energy systems in specific locations choose to build additional flexibility into their regulations to accommodate solar. For instance, some communities allow solar installations of less than a certain height to be placed in the setbacks of an individual lot. Other communities allow solar structures to be exempted from setback, height, and lot coverage restriction in certain districts. Take each of these considerations into account as part of the review and assessment process to determine the most suitable dimensional standards and level of flexibility for each zoning district within your community.

Development Standards

Development standards address site and building design considerations unique to a specific use or location within the community. Depending on the community, development standards may be located within the zoning code, the subdivision regulations, or addressed in both locations. With respect to solar energy use, development standards typically address issues such as screening, fencing and enclosures, and the protection of solar access.

Typical Regulatory Issues. Regulatory issues with these standards tend to center either on a lack of standards or on the difficulty of administering existing standards. Without basic development standards in place to guide the installation of solar energy systems, potential conflicts between uses may arise. While conflicts can certainly arise due to a lack of screening or fencing requirements, the potential for conflict is most likely to occur where standards do not exist to protect solar access. Without solar access standards in place, the existing and future solar potential of a particular site may be significantly reduced or eliminated altogether due to buildings, trees, or other potential obstructions. However, when solar access standards do exist, they can be extremely challenging to enforce and administer—either because they are too general or because they are too restrictive.

Key Questions.
- Does the code include development standards to minimize the potential impacts of solar energy systems on adjacent uses and address safety and environmental concerns (e.g., screening or fencing requirements for solar energy systems that function as a primary use and limits on the amount of glare that can reasonably be transmitted to adjoining properties)?
- Does the code contain development standards to maximize or protect solar access?
- Does the code establish parameters to guide the resolution of potential conflicts that arise in the process of enforcing solar access requirements?

- Does the code establish a process through which the removal of nonfunctional or decommissioned systems can be enforced?

Special Considerations. Solar access requirements are often not practical citywide, particularly in urban locations or where higher density development is desired. Focus requirements where resources are most viable or development patterns are most conducive to supporting your community's solar objectives. When developing new standards, take into account the administration of solar access requirements over time. If your community has limited staffing resources and solar expertise, avoid adopting detailed solar access requirements that are likely to be more onerous to administer.

Communities wishing to go above and beyond baseline regulations for solar may choose to include one or both of the following options as part of their package of solar regulations:

- *Solar site design requirements*—These standards establish basic site design parameters for "solar-oriented lots" that are intended to increase solar access to individual lots and preserve future options for solar. They typically require streets and lots to be oriented to maximize solar access and may also allow for flexible setbacks to accomplish the same objective. They typically apply to new subdivisions and are imposed without any degree of certainty that solar energy systems will be installed in the future. Solar siting requirements may be limited to single-family residential zone districts or other lower-density districts or applied to residential dwellings in all zones. To provide flexibility, these requirements stipulate that some percentage of the total lots in a particular development must comply with the site design standards. Percentages may range from as little as 30 percent to more than 80 percent of lots. It is important to account for topography, tree canopy, and other site considerations when developing such requirements for a particular community to ensure the standards are not so restrictive as to effectively prohibit development altogether.
- *Solar-ready home requirements*—To encourage the broader use of solar over time, some communities require new homes to be "solar ready." These requirements are typically incorporated as part of the building and or plumbing code and often include structural/roof specifications, solar "stub-in" requirements for new homes to support future photovoltaic panel or hot water heater installation, and installation of PV conduit or hot water pipes on south, east, or west-facing roofs.

Definitions

Every code includes a list of defined terms. These definitions are intended to reduce the need for interpretation in the administration of the code and to minimize inconsistencies in the review process.

Typical Regulatory Issues. If a code is silent on the issue of solar energy systems and related concepts in its list of defined terms, applicants and planners must interpret how these systems relate to terms that are defined, which may lead to delay or denial of solar applications during the review process.

Key Questions.

- Does the code define the following key terms associated with solar energy systems?
  - Solar access
  - Solar collectors
  - Solar energy systems (with separate definitions for large vs. small and building vs. ground-mounted systems, as applicable)
  - Solar garden
  - Solar farm
  - Solar-ready buildings
- If the code defines solar-related terms, do these terms reflect current solar technologies and practices? If not, do they need to be revised or replaced as part developing new solar regulations?

Special Considerations. Defined terms should be tailored to prevalent technologies being used locally. In addition, they should be aligned with solar terminology used in other codes throughout the region, if applicable. For example, some communities protect solar access by regulating the "solar envelope" of a building—others define it as a "solar fence." The use of consistent terminology among communities in a region reduces potential for confusion and delay and simplifies the application process for all parties involved. Local solar installers can often serve as a resource to help identify a comprehensive list of terms that should be included and possible conflicts between solar regulations in neighboring communities. When drafting new definitions, ensure that the language is specific enough to avoid difficult interpretations, but general enough to accommodate changes in solar technologies over time.

STEP 2: DEVELOP SOLAR REGULATIONS

The second step in the process is to build on the information gathered as part of the review and assessment phase to develop an initial draft of the actual solar regulations. Baseline and optional provisions for solar regulations are summarized below. Specific direction on each of these considerations should be drawn from the review and assessment summary report, discussions with community stakeholders, as well as the many model and sample regulations referenced in these appendices.

Baseline Considerations

To ensure solar requirements are clear, comprehensive, and enforceable, it is important to address each of the baseline considerations described below.

Purpose and Intent. Clearly describe the purpose and intent of the solar development regulations using existing plans and policies and the results of the review and assessment process as a guide. This purpose statement may range from a straightforward desire to encourage solar energy use communitywide to a comprehensive list of objectives that stem from a desire to clarify how and where solar development is encouraged or discouraged within the community.

Applicability. Define the types of development that the solar standards will apply to and where. In most instances, solar development regulations will apply to all types of solar energy systems and will apply citywide. If the baseline considerations are being addressed through a series of targeted amendments to an existing code (e.g., minor revisions or additions to various sections of the code), it may not be necessary to include a statement of applicability; however, such a statement can be helpful to include in interim drafts of the solar standards to provide context for the proposed amendments.

Permitted Uses. Define what types of solar energy systems will be allowed in which zoning districts as a primary use or accessory use, or with a special use permit. If solar energy systems will be prohibited in certain districts, clearly distinguish those districts. Provide cross references to use-specific dimensional and development standards, as necessary.

Dimensional Standards. Define the basic dimensional standards—maximum height, minimum setbacks, and maximum lot coverage—that will apply to different types and sizes of solar energy systems in different zone districts (e.g., standards for a ground-mounted solar energy system in an industrial district are likely to be less restrictive than those for a roof-mounted solar energy system in a historic district). If flexibility in the application of these standards will be allowed for solar energy systems, clearly define where that flexibility is allowed and what the process is for taking advantage of that flexibility (e.g., systems may be placed within side yard setbacks by right).

Development Standards. Establish specific development standards that will apply to different types and sizes of solar energy systems in different zone districts (e.g., screening requirements for freestanding systems or glare restrictions for hillside systems). If a particular standard or set of standards will be applicable to solar citywide (e.g., removal of non-functional or decommissioned systems), consider grouping those standards under a "general" category to reduce repetition and complexity in the solar regulations.

Definitions. Ensure all key terms used in the solar development regulations are clearly defined, yet are drafted broadly enough to allow for changes in technology over time. In particular, focus on specific types of solar energy systems and site considerations. Cross-check lists of terms with those used by adjacent communities for consistency.

Optional Provisions

If appropriate, incorporate the following optional provisions within the overall package of solar requirements. Alternatively, some communities may choose to pursue one or both of these options after their baseline solar requirements have been adopted and in place for a period of time. This "waiting period" allows a community to apply the baseline solar requirements in practice, providing a clearer understanding of staff time needed for administration, the volume of applications likely to occur on a weekly or monthly basis, and the overall level of support for more robust solar requirements.

Solar Site Design. Establish specific standards to promote solar access. In particular, take care to consider the appropriate level of flexibility for solar energy systems in different zone districts within the community based on their predominant land-use type and density (both existing and planned), existing tree canopy, topographic characteristics, and the degree to which the area has been "built out."

Solar-Ready Home Requirements. Work closely with building department officials and local installers to develop required specifications for solar-ready homes and to determine where the requirements would be most beneficial over time. Typically, such requirements are most effective in an area likely to see significant new development.

CONCLUSION

The act of establishing solar development regulations for a community is an important step toward reinforcing community support for solar energy use, identifying and removing potential barriers to using the solar resource, and minimizing the potential impacts of different types of solar energy systems. At a minimum, solar regulations should address what types and sizes of solar energy systems are allowed and where, mitigate potential compatibility issues associated with solar equipment, and define and protect solar access. Optional considerations may include more robust solar access, solar site design, or solar-ready homes requirements. Regardless of the approach selected, solar development regulations must be carefully tailored to meet the unique size, geography, climate, regulatory framework, and political/natural environment of individual communities. The model ordinance framework presented in this appendix provides a guide for refining existing regulations to more clearly address solar energy systems or to develop a stand-alone solar ordinance from scratch.

Model Solar Development Ordinances

California County Planning Directors Association
Model SEF Permit Streamlining Ordinance; Renewable Energy Combining Zone
http://www.ccpda.org/solar

- Model permit streamlining ordinance establishes four tiers of solar energy facilities and four use categories (direct use, accessory use, secondary use, and primary use). Small ground-mounted and all rooftop systems are permitted by right as accessory uses; larger ground-mounted systems require administrative permits, minor use permits, or conditional use permits depending on their intensity and location. Standards address height limits, setbacks, abandonment, erosion control, visibility in scenic areas, and protection of agricultural and biological resources.

- Model renewable energy combining district establishes overlay for large-scale renewable energy facilities within nonprime agricultural lands as well as resource, general commercial, heavy industrial, and public facilities districts. Solar energy facilities up to 30 acres in area and associated transmission lines less than 100 kV and substations are permitted with an administrative permit. Larger sites and transmission lines require a use permit. Development standards address aesthetics, air quality, air safety, biological resources, cultural and historic resources, agricultural resources, erosion and sediment control, fire protection, grading, security, signs, decommissioning and restoration, financial assurance, and workforce development.

Columbia Law School, Center for Climate Change Law
Model Small-Scale Solar Siting Ordinance (2012)
Danielle Sugarman
http://web.law.columbia.edu/climate-change/resources/model-ordinances/model-small-scale-solar-siting-ordinance

- Model ordinance for small-scale solar energy systems (solar thermal and PV systems up to 10 kW that do not send energy off site) includes purpose provisions and extensive definitions section.

- Permits small-scale solar energy systems by right in all districts with building permit and subject to conditions. Includes provisions for rooftop- and building-mounted systems, building-integrated PV systems, ground-mounted solar collectors, and solar thermal systems. Includes safety provisions.

- Optional provisions address solar fast-track permitting program, zoning for future solar access, and tree removal.

Cumberland County, Pennsylvania, Planning Department
Solar Energy Systems Model Ordinance
http://www.ccpa.net/DocumentCenter/Home/View/7947

- Provides definitions for a range of solar terms.

- Provides standards for accessory solar energy systems; provisions address certified installers, glare, decommissioning, and solar easements. Provides additional, separate standards for roof- and wall-mounted systems as well as ground-mounted systems.

- Provides standards for principal solar energy systems; provisions address certified installers, glare, decommissioning, solar easements, impervious surfaces, security, access. Provides additional, separate standards for ground-mounted systems as well as roof- and wall-mounted systems.

Delaware Valley Regional Planning Commission, Alternative Energy Ordinance Working Group
Renewable Energy Ordinances Framework—Solar
http://www.dvrpc.org/EnergyClimate/ModelOrdinance/solar.htm

- Model offers solar ordinance framework with suggested language options and commentary. Framework covers purpose, definitions, applicability, and general regulations (height and setbacks, aesthetics and screening, solar access, first responder safety, and compliance with other regulations).

Kent County, Maryland
Renewable Energy Task Force White Paper
Solar Energy Systems: Proposed Land Use Ordinance Language
http://www.kentcounty.com/gov/planzone/RETF_WHITE_PAPER_Final.pdf

- Defines "utility scale" (energy used offsite) and "small" (energy primarily used on site) solar energy systems.
- Permits utility-scale solar energy systems by right in industrial and employment center districts and as special exception uses in agricultural and commercial districts with conditions.
- Permits small solar energy systems in commercial districts by right and as accessory uses by right in agricultural and residential districts with conditions.

Lancaster County, Pennsylvania, Planning Commission
Municipal Guide to Planning for and Regulating Alternative Energy Systems (2010)
http://www.co.lancaster.pa.us/toolbox/lib/toolbox/alternativeenergyguide/alternative_energy_guide_pdf.pdf

- Model ordinance for alternative energy systems includes definitions for "accessory solar energy system" and "principal solar energy production facility."
- Provides model ordinance language permitting accessory solar energy systems in all districts subject to standards listed in ordinance; standards address applicability, design and installation, height restrictions, setbacks, plan approvals, utility notification, and limitations on solar energy systems restrictions.
- Provides model ordinance language permitting solar energy production facilities in industrial and commercial zones by right and in agricultural districts by special exception. Includes standards covering design and installation and decommissioning, among other things (Oregon DOE model language).

Mid-America Regional Council, Solar Ready KC
Best Management Practices for Solar Installation Policy: Planning Improvements Step 1B: Improve Solar Access
http://www.marc.org/environment/energy/assets/BMP%20Planning%20Step%201-1%20B%20Improve%20Solar%20Access.pdf

- BMP sheet offers model ordinance to encourage access to solar energy, which addresses solar easements and solar rights, and provides sample ordinance language for purpose statements, solar zones, solar fences, solar siting, and solar access permits.

Minnesota Environmental Quality Board
From Policy to Reality: Updated Model Ordinances for Sustainable Development (2013 revision)
Brian Ross, CR Planning
Solar Energy Standards
http://www.crplanning.com/_ordinances/solar.pdf

- Model provisions include purpose statement linking regulations to comprehensive plan goals and other community values.
- Includes extensive definitions section.
- Allows solar energy systems as permitted uses in all districts subject to standards addressing height, setbacks, visibility, plan approval, and code compliance. Allows for administrative variance or conditional use permit for situations where standards cannot be met.

- Prohibits private covenants from restricting the use of solar energy systems.
- Provides for solar access easements.
- Provides that communities may require solar energy systems installation as a condition for rezoning, conditional use permits, or PUD approval.
- Offers a range of incentives for solar roofs.

Monroe County, Pennsylvania

Model Ordinance for On-Site Usage of Solar Energy Systems

http://www.co.monroe.pa.us/planning_records/lib/planning_records/planning/model_monroe_county_on-site_usage_of_solar_energy_systems.pdf

- Permits solar energy systems by right as accessory uses in any zoning district subject to listed criteria addressing height, setbacks, screening, glare, permitting, design, and ground-mounted system decommissioning.

Oregon Department of Energy

A Model Ordinance for Energy Projects (2005)

http://www.oregon.gov/energy/Siting/docs/ModelEnergyOrdinance.pdf

- Model ordinance for large-scale energy facilities provides general siting and permitting standards as well as specific standards for different types of energy sources, including solar. Standards for solar facilities address protection of natural ground contour and wildlife resources, prohibit glare, and require public safety plan and minimal use of hazardous chemicals or solvents.

PennFuture, Western Pennsylvania Rooftop Solar Challenge

Final Solar Zoning Ordinance (2012)

http://www.pennfuture.org/SunShot/SunSHOT_Ord_Zoning.pdf

[required online form for download]

- Model ordinance intended for use by range of western Pennsylvania municipalities provides for approval by right of building- and ground-mounted solar energy systems (50 kW or less at residential sites, 3,000 kW or less at other sites) as accessory uses in all zoning districts. Standards address location within a parcel, design and installation standards, setbacks and height, screening and visibility, impervious coverage, inspections, and nonconformities.

Solar ABCs, Colleen Kettles

A Comprehensive Review of Solar Access Law in the United States (2008)

www.solarabcs.org/about/publications/reports/solar-access/pdfs/Solaraccess-full.pdf

Model Statute/Ordinance to Encourage Access to Solar Energy

- Model ordinance provides for creation of solar easement by covenant or solar access permit, lists required content of easement instrument, and prohibits solar-restrictive private covenants.

SolarTech, Troy A. Rule

Legislating For Solar Access: A Guide and Model Ordinance (2012)

http://solar30.org/wp-content/uploads/2013/04/Guide-to-Solar-Access-Ordinance.pdf
http://solar30.org/wp-content/uploads/2013/04/Model-Solar-Access-Ordinance.pdf

- Model solar access ordinance establishes solar access overlay zone in which property owners may apply for solar access easement. Lists criteria for solar access regulatory board to consider in deciding to approve the easement; requires compensation of at least $500 to be paid to affected adjacent property owners. Provides for removal of easement.

Virginia Department of Environmental Quality

Model Ordinance for Smaller-Scale Solar Energy Projects in Virginia (By Right Permitting)(2012)
Model Ordinance for Larger-Scale Solar Energy Projects in Virginia (2012)
Model Ordinance: Solar Tax Exemption in Virginia (2012)

http://www.deq.virginia.gov/Programs/RenewableEnergy/ModelOrdinances.aspx

- Model ordinance with commentary for smaller-scale solar energy projects (projects smaller than or equal to 2 acres or installed on or over buildings, parking lots, or previously disturbed areas) provides for administrative plan approval for these systems in all districts if standards are met. Standards address location, appearance, and operation, as well as decommissioning.
- Model ordinance with commentary for larger-scale solar energy projects (projects larger than 2 acres or not installed on or over buildings, parking lots, or previously disturbed areas) allows for approval by right in agricultural and industrial zones and requires special use approval in residential and commercial zones. Lays out application and approval process procedures and requirements; standards address location, appearance, and operation of project sites, as well as decommissioning.
- Model tax exemption ordinance enables localities to exempt solar energy equipment and facilities from local property taxes.

Wasatch [Utah] Solar Challenge
Model Ordinance for Residential and Non-Residential Distributed Solar Energy Systems
http://solarsimplified.org/zoning/solar-zoning-toolbox/solarzoningordinance

- Model ordinance permits solar energy systems in any districts as accessory uses, subject to specific criteria. Includes extensive definitions section (45 terms); provides development standards for roof-mount and ground-mount systems; addresses safety and abandonment/removal. Additional provisions address solar-ready zoning and prohibit private covenants that restrict the use of solar collectors.

Solar-Supportive Development Regulations

State	Community	Municipal Code Citation	Accessory-Use Solar Energy System Standards	Principal-Use Solar Energy System Standards	Community Solar System Standards	Solar Access Protections	Solar Site Design	Solar-Ready Homes	Competing Priorities	Incentives
Alabama	Huntsville, City of	Ordinance 12–466	x	x						
Alaska	Anchorage, City of	§21.40.150.H.2								x
	North Slope Borough	§18.20.130.F					x			
Arizona	Chandler, City of	§35-2210	x	x						
	Oro Valley, Town of	§27.1.A; Ordinance No. (O)09-11				x		x		
	Phoenix, City of	§1223.C		x						x
	Pima County	§18.07.030.P	x							
	Pinal County	§2.210.010 et seq.	x							
	Tuscon, City of	§3.2.5.2.E; §3.2.12; §3.6.1.6.C; Ordinance No. 10549	x			x	x	x		
Arkansas	Bentonville, City of	Zoning §601.24	x							
California	Butte County	§24-157	x	x						
	Calimesa, City of	§18.20.060.C	x			x	x	x	x	
	Chico, City of	§19.60.100; §19.37.120.A.9	x						x	
	Del Mar, City of	Ch.23.20; Ch.23.51	x			x			x	
	Lancaster, City of	§17.08.270 et seq.	x	x				x		
	Rancho Palos Verdes, City of	§15.04.070; §17.83.050;	x							
	San Luis Obispo, City of	§16.18.170, §16.17.080.C.g.				x		x		x
	Santa Barbara, City of	§28.11.010 et seq.				x				
	Santa Clara County	§C12-173 et seq.; §1.4.40.020.M; §1.4.10.345	x	x		x	x			
	Taft, City of	§6.11.330	x				x			
Colorado	Aurora, City of	§§146-1280-83	x	x						
	Boulder, City of	§9-7-7, §9-9-17	x			x	x			
	Carbondale, City of	§15.30.130; §17.96.010 et seq.; §18.50.010.D		x		x	x	x		
	Fort Collins, City of	§3.2.3	x			x	x		x	
	Longmont, City of	§16.03.130.H					x			x

State	Community	Municipal Code Citation	Accessory-Use Solar Energy System Standards	Principal-Use Solar Energy System Standards	Community Solar System Standards	Solar Access Protections	Solar Site Design	Solar-Ready Homes	Competing Priorities	Incentives
Connecticut	Enfield, Town of	§46-83; Zoning §8.80	x	x					x	
	Farmington, Town of	§111-26.A							x	x
	Haddam, Town of	§302-36				x	x		x	x
	Meriden, City of	§213-53.B	x			x				
	Naugatuck, Borough of	Subdivisions §4.3-4, §4.4.2, §4.16, §4.19					x		x	
	Orange, Town of	§382-30; §382-23.O				x	x		x	
Delaware	Bethany Beach, Town of	§484-1 et seq.	x							
	Fenwick Island, Town of	§160-9.A	x							
	Henlopen Acres, Town of	§43-5.J	x							
	Lewes, City of	§197-59.11							x	
	Newark, City of	Subdivisions App. XI §I					x			
Florida	Boynton Beach, City of	LDR Chap. 3, Art. V, §3.W; Chap. 4, Art. IX, §6.H.1A.q.6	x							
	Broward County	§39-109	x						x	
	Pinecrest, Village of	LDR Art. 5, Div. 5.27	x					x	x	
	St. Lucie County	§7.10.28; §8.00.03.L	x	x						
	Tamarac, City of	§24-615	x						x	
Georgia	Norcross, City of	§§115-38.e-f	x							
	Putnam County	§66-72.a; §66-112.a		x						
Idaho	Blaine County	§9-3B-3; §10-5-3.N.4.g; §10-6-6.D	x				x			x
	Buhl, City of	§9-24D-13.D				x				
	Gem County	§11-6-5.M		x						
	Ketchum, City of	§16.04.040.F; §17.96.090.B.6					x			
	McCall, City of	§3.10.024.A								x
	Pocatello, City of	§16.32.070; §16.32.080					x			
	Valley County	§9-3-1; §9-5G-1	x							

State	Community	Municipal Code Citation	Accessory-Use Solar Energy System Standards	Principal-Use Solar Energy System Standards	Community Solar System Standards	Solar Access Protections	Solar Site Design	Solar-Ready Homes	Competing Priorities	Incentives
Illinois	Buffalo Grove, Village of	§17.28.050.E.3.i								x
	Grundy County	§8-2-4-11; §8-2-5-30	x	x		x				
	New Lenox, Village of	§106-501 et seq.	x							
	North Aurora, Village of	Zoning, §12.3.I	x							
	Roselle, Village of	Zoning, §3.02; §7.01.C				x	x			
	Schaumburg, Village of	§154.56, §154.59; §151.06.F	x			x	x			
	Sugar Grove, Village of	§11-4-21	x							
	Will County	§155-10.10.G	x							
	Woodridge, Village of	§9-12-7	x							
Indiana	Beverly Shores, Town of	§155-135				x				
	Boone County	§157.034.C.3.c							x	
Iowa	Ames, City of	§29.1309	x							
	Cedar Rapids, City of	§32.05.010.4.f	x							
	Dubuque, City of	§14-14-1 et seq; §16-5-2-3 et al; §16-11-12					x			
	Mason City, City of	§12-8-2 et al.; §12-21-1 et seq.	x			x				
	Pella, City of	§165.26.5	x							
	Spencer, City of	§9-11-8.D.4	x							
	Story County	§85.08.198; §86.04.3.E et al.; §86.16.6	x				x			
Kansas	Greensburg, City of	Zoning §4.1	x			x				
	Independence, City of	Zoning §1501				x				
	Lenexa, City of	§4-1-B-24-F.12	x							
	Lindsborg, City of	§§51-133.d-e				x	x			
	Olathe, City of	§18.56.110.L	x							
	Overland Park, City of	§18.390.140.K	x							
Kentucky	Lexington-Fayette Urban County	Zoning §15-1.c.4	x							
Louisiana	Addis, Town of	Zoning §6					x			
	Gretna, City of	§52-9.6							x	

State	Community	Municipal Code Citation	Accessory-Use Solar Energy System Standards	Principal-Use Solar Energy System Standards	Community Solar System Standards	Solar Access Protections	Solar Site Design	Solar-Ready Homes	Competing Priorities	Incentives
Maine	Belfast, City of	§90-42.b.17				x				
	Old Orchard Beach, Town of	§74-273					x			
	Shapleigh, Town of	§89-36.C; §105-54	x				x			
	South Berwick, Town of	§121-25					x			
Maryland	Berlin, Town of	§108-274					x			
	Denton, Town of	§128 App. 2 §2.B							x	
	Dorchester County	§155-50.LL; §155 Att 1		x						
	Hagerstown, City of	§140-32.L	x							
	Laurel, City of	§20-20.8	x							
	Middletown, Town of	§17.38.010 et seq.; §17.48.400	x							
	Queen Anne's County	§18:1-95.S		x						
	Worcester County	§ZS1-44	x	x						
Massachusetts	Belchertown, Town of	§145-28		x						
	Holyoke, City of	§18-99.c; Zoning §7-9	x	x					x	
	Bellingham, Town of	§240-162 et seq.		x						
	Orange, Town of	§205.3340 et seq.					x			
	Plainville, Town of	§500-26.1		x						
Michigan	Bay City, City of	§122-621 et seq.	x	x						
	Canton, Charter Township of	Zoning §6.04.A.4					x		x	
	Casco, Township of	§§13.24-26	x	x	x				x	
	Ferndale, City of	§24-183.3.b	x							
	Grand Rapids, City of	§5.11.14.B	x							
	Greenville, City of	§46-73	x							
	Royal Oak, City of	§770-54	x							
	Tecumseh, City of	§§98-552-53	x			x				

State	Community	Municipal Code Citation	Accessory-Use Solar Energy System Standards	Principal-Use Solar Energy System Standards	Community Solar System Standards	Solar Access Protections	Solar Site Design	Solar-Ready Homes	Competing Priorities	Incentives
Minnesota	Brown County	§702	x			x				
	Cottage Grove, City of	§10-4-4; §11-9E-6; §11-10B-1.C	x			x	x			x
	Houston County	Zoning §0110.2808	x			x				
	Faribault, City of	UDR §6.230.G	x			x				
	Kellogg, City of	§220-56	x							
	Minneapolis, City of	§535.820 et seq.; §551.850.2; §598.240.3	x			x	x		x	
	Stearns County	Zoning §6.51	x	x						
	Watertown, City of	§61-13 et seq.	x							
Mississippi	Biloxi, City of	§23-6-12.B.2; §23-4-4.B; §23-4-4.C.21	x							x
Missouri	Clay County	§151-6.3.G	x							
	Gladstone, City of	§7.167.030	x							
	Kansas City, City of	§88-305-09	x							
	Perryville, City of	§17.56.01.F				x				
	St. Peters, City of	§§405.536.B-C	x							
	Warrensburg, City of	§27.240	x	x						
Montana	Great Falls, City of	§17.28.050, Exh. 28-1.3					x			
	Butte-Silverbow County	§17.36.090	x							
Nebraska	Alliance, City of	§115-111.b.4.e; §115-170.e.8	x							x
	Bellevue, City of	Zoning §8.06	x							
	Gothenburg, City of	§151.054					x			
Nevada	Churchill County	§16.08.250; §16.08.260; §16.16.030	x	x						
	Henderson, City of	§19.5.7.D.10; §19.7.6.D.h.3; §19.7.12.C	x			x	x			
	Las Vegas, City of	UDC §19.10.150.O; §19.12.070	x					x	x	x
	Sparks, City of	§20.103.040	x	x						
New Hampshire	Francestown, Town of	Zoning §7.18	x	x		x				
	Gorham, Town of	Zoning §§4.01A–B et al.	x							
	Hampton, Town of	Zoning Art. XVIII	x							

State	Community	Municipal Code Citation	Accessory-Use Solar Energy System Standards	Principal-Use Solar Energy System Standards	Community Solar System Standards	Solar Access Protections	Solar Site Design	Solar-Ready Homes	Competing Priorities	Incentives
New Jersey	Bethlehem, Township of	§102-37.3	x	x						
	Glassboro, Borough of	§107-72	x							
	Harmony, Township of	§165-10.C.2 et al.; §165-45.1	x	x						
	Hoboken, City of	§196-35.1	x						x	
	Holland, Township of	§100-20.1; §100-21.M	x	x						
	Montgomery, Township of	§16-5.5.e; §16-6.10	x	x			x			
	Wall, Township of	§140-139.1		x						
	Washington, Township of (Burlington County)	§275-85				x	x			
New Mexico	Albuquerque, City of	§14-11-1 et seq; §14-14-4-2.B; §14-14-4-7; §14-16-2-11.C				x	x			
	Angel Fire, Village of	§7-1-4-1.C; §9-11C-5.K	x			x	x			
	Deming, City of	§12-4H-7; §12-4I-6				x	x			
	Los Alamos County	§16-279				x				
	Ruidoso, Village of	§§54-140.2, 4	x						x	
	Taos, Town of	§15.16.010; §16.16.220.8.D.1.6; §16.20.030.1.G.5	x			x			x	
New York	Albany, City of	§375-93	x							
	Big Flats, Town of	§17.36.140				x	x			
	Brookhaven, Town of	§§85-556-61	x	x						
	Elmira, Town of	§217-73				x	x			
	Ithaca, Town of	§234-25; §270-219.1	x			x	x			
	Mastic Beach, Village of	§415-2	x							
	Westfield, Village of	§155-57				x			x	
North Carolina	Brunswick County	UDO §5.3.4.P; §5.4.10	x	x						
	Camden County	§151.334; §§151.347.V-W	x	x						
	Chapel Hill, Town of	Land Use §3.8				x				
	Granville County	§32-142; §32-162.5; §32-163.7; §32-233		x						
	Kure Beach, Town of	§19-341	x							
	Pleasant Garden, Town of	DO Table 4-3-1; §6-4.69	x	x						
	Selma, Town of	§17-127.c.7		x						